W9-DHG-161

The

VELOCITY
MANIFESTO

Harnessing Technology, Vision, and Culture
to Future-Proof your Organization

SCOTT KLOSOSKY

Greenleaf
Book Group Press

Published by Greenleaf Book Group Press
Austin, Texas
www.gbgpress.com

Distributed by Greenleaf Book Group LLC

For ordering information or special discounts for bulk purchases, please contact Greenleaf Book Group LLC at PO Box 91869, Austin, TX 78709, 512.891.6100.

Design and composition by Greenleaf Book Group LLC and Bumpy Design
Cover design by Greenleaf Book Group LLC

Publisher's Cataloging-In-Publication Data
(Prepared by The Donohue Group, Inc.)
Klososky, Scott.   The velocity manifesto : harnessing technology, vision, and culture to future-proof your organization / Scott Klososky. — 1st ed.   p. ; cm.
  ISBN: 978-1-60832-085-1
  1. Technological innovations--Management. 2. Business enterprises—Technological innovations. 3. Information technology—Management. 4. Strategic planning.
5. Success in business. I. Title.
HD58.8 .K55 2011
658.406   2010939240

Part of the Tree Neutral® program, which offsets the number of trees consumed in the production and printing of this book by taking proactive steps, such as planting trees in direct proportion to the number of trees used: www.treeneutral.com

Printed in the United States of America on acid-free paper

11 12 13 14 15   10 9 8 7 6 5 4 3 2 1

First Edition

No one learns to be a leader as a solo effort, nor does anyone write a book without the help of many others. If I were to attempt to thank everyone who has helped me with either of these pursuits, pages would be filled with names, and that would bore you. I will have to find other ways to repay the kindness of the people that have invested in me over the years. It is my hope that this book will play at least a small role in passing on what I have learned so that you may benefit, as I have, by standing on the shoulders of others. There is one special person to whom I must dedicate this book, because she has always been dedicated to me: my wife, Annette. I would not have the capacity to be the scribe without her support.

CONTENTS

INTRODUCTION

A rock pile ceases to be a rock pile the moment a single person contemplates it, bearing within them the image of a cathedral.

—Antoine de Saint-Exupéry

MANY OF YOU WILL LOSE YOUR LEADERSHIP POSITION in five years or less unless you embrace the velocity manifesto. This is a fact. The average tenure of a CEO in the United States is 4.2 years, less than half the average tenure of 10.5 years in 1990, according to Harvard Business School professor Michael Beer. If that trend continues, we will soon be down to a "Leader for a Year" model.

For more than a decade, I have been helping people like you adapt to the technological innovations that are changing the way we live and do business so that they can retain—and succeed in—their management positions. You can no longer afford to ignore the impact of technology on your current operations, your future operations, and your organizational culture. Showing this truth to today's leaders is the essence of the velocity manifesto.

You want the pithy version? Keep pace or go home.

This book will show you how to keep pace as a leader in a marketplace that moves at scorching speed. You will learn concepts, processes, and leadership skills that will enable you to increase the velocity of your organization and thrive in today's technology-driven environment. In these pages, I will discuss everything from your company's digital infrastructure to your work-at-home policy and show you how to take immediate action to improve your rate of success. So hang on tight.

The pace of your career, and of your life, is speeding up, and I know you can sense it. Technological innovations are faster, product development cycles are shorter, and expectations for everything are higher. In short, the velocity of business, driven by technological change, is increasing exponentially. This increased velocity demands new leadership skills, skills that are very different from the ones that were effective twenty years ago. Although some leadership abilities are timeless, learning these new, high-velocity leadership skills is absolutely essential for any leader in any organization.

I spend most of my days either speaking to leaders or consulting inside their organizations. I have seen leaders who understand the velocity manifesto, and leaders who don't. Sometimes age is a factor, but not always. Leaders in conservative industries like banking, manufacturing, accounting, and law are more likely to be immobilized by inertia rather than to adapt to the new, high-velocity environment. Many of these leaders have wonderful attributes but are missing critical skills that are just plain mandatory today. In the most tragic cases, people who were on top of their game just ten years ago are now struggling to figure out how to make decisions in a fast-paced world that is foreign to them.

My experiences as an entrepreneur and adviser have inspired me to write *The Velocity Manifesto*. I feel blessed to have been able to help clients learn new vocabulary, concepts, and processes to upgrade their inventory of skills. The start of my career coincided with the advent of the PC (personal computer), and I have spent decades building technology companies. This background has forced me to keep pace with the growing velocity of business. I want you to benefit from my experience and see the world through my eyes; it's frustrating that so many leaders are unwilling to learn new skills and are simply fading away along with their organizations. I want you to understand that your survival depends on the concepts presented in this book.

The Velocity Manifesto will help you see the world from new perspectives and take huge steps toward the goals you want your organization to accomplish. That is, after all, what being a leader is all about.

The Velocity Manifesto addresses three key areas in which you must develop an expertise:

1. **Digital Plumbing:** Building an efficient and productive technology infrastructure that gives you an advantage over the competition.

2. **High-Beam Strategy:** Setting an accurate vision and direction for the future of your organization.

3. **High-Velocity Culture:** Creating a highly productive workforce that integrates different types of skills and lifestyles.

Fail in the first area, and your organization will not be able to compete with companies that have done a better job than you at technology implementation.

Fail in the second area, and your organization will wander around, lost.

Fail in the third area, and you will not have the team you need to accomplish your organization's goals.

Each of the three sections of this book will show you how to develop specific leadership skills. The first section lays out new concepts and processes you can use to leverage technology—or what I refer to as your "digital plumbing"—as a tool for progress. It is written for the "civilian" leader who does not want to be a geek but needs to understand how to apply technology holistically. The ideas in it are not trendy or short-lived; they are universal truths about technology that will enable you and your organization to keep up with the velocity of technological change.

The second section shows you how to more accurately predict the future by learning how to spot trends and take advantage of them with a portfolio of targeted investments. As the pace of change increases, your organization has to be able to see further into the future—using what I call a "high-beam strategy"—in order to leverage trends and mitigate risks. Once you have built a sophisticated digital plumbing system as a foundation for growth, you need this strategy to know how to put it to work. A powerful tool is worthless if the wielder has no vision.

The last section discusses the cultural changes that drive organizations today. The workforce is evolving, and you need to make conscious adjustments to your organization in order to diminish friction among your employees. You also must know how to manage a transitional, multigenerational workforce that is characterized by conflicting views on how to live and how to get things done. Plus, people are working more and more on virtual teams that must be organized to leverage our new, multimillion-dollar digital infrastructures. Without effective teams, a powerful digital plumbing platform and a wonderfully accurate vision of the future guarantee nothing. If your team cannot take full advantage of technology with gusto and verve, they will lie fallow—and you will fail as a leader.

When you add up all your leadership skills, do you believe that you are ready to guide your organization as it renovates its digital plumbing, develops a high-beam strategy, and migrates to an effective team culture? If you have any doubt about it, read on. And if you think you have all this down cold, then pass this book to someone else in the organization who might benefit from it.

One final note before we get started: *The Velocity Manifesto* is not an intellectual exercise. It's an action plan for you to apply—and apply hard. Too many of us think that the job of changing our organizations, teams, or the world should be left up to someone else. Too many of us sit around and wonder, "Who am I to think I can lead?" The reality is that you are *the* single person who can lead in your corner of the world, and you must set a pace for your organization that matches the pace of the marketplace. Like I said, keep pace or go home. I don't intend to let any of you go home.

Get ready for a wild ride. I have no intention of presenting you with one simple idea and then explaining it to death. That's not the way I work. Instead, you will be introduced to a wide variety of concepts and processes that will affect everything from your firewalls to your strategy for the future and maybe even the way you do organizational charts. I will, of course, show you how to drill down and get more complete information on any of the processes I discuss, but my goal here is to open up the fire hose and hit you with a steady stream of new ideas. Any leader can benefit from adding a few new tricks to the inventory, and it is my hope that *The Velocity Manifesto* will shake you up and change the way you lead your organization. We've got a lot of ground to cover, so let's get started.

Note: As you read this book, you will see an icon that looks like this to the side of paragraphs where we discuss additional resources that can be found on the website for this book (www. velocitymanifesto.com). In many cases, there are process documents, web links, or white papers that further explain a topic we have touched on. The website also provides a way for you to contact me about any further questions you might have or just to register an opinion on the book. I would love to hear what you think about the concepts included here.

PART ONE

RENOVATING YOUR DIGITAL PLUMBING

CHAPTER 1

EFFICIENT TECHNOLOGY = VELOCITY

The illiterate of the 21st century will not be those who cannot read and write, but those who cannot learn, unlearn, and relearn.

—ALVIN TOFFLER

IF YOU STOP AND THINK ABOUT IT, YOU'LL REALIZE that nothing has changed the dynamics of how an organization operates more than technology. No other variable can decimate existing structures and create changes in markets at the lightning speed that technology does. In fact, no other man-made variable will have more impact on our lives. Even the poorest people in the poorest countries are beginning to benefit from advances in medical, transportation, and communication technologies. People who can barely find enough food to eat often have a cell phone in hand. In the developed world, we are quickly integrating technology into every minute of every day. We depend on it to stay healthy, to earn a living, to communicate, and to prosper both as human beings and financially.

And as it does all this, technology is also reorganizing how we operate organizations.

Technology is the main catalyst behind the increasing velocity of change in our world. It allows us to replace many human tasks with automated systems that perform much faster and more accurately. This has been true since the invention of the wheel, which helped humans increase the velocity of their transportation systems and cut down on the number of people needed to move products. The speed of technological innovation has been increasing over the centuries, but it took off exponentially with the advent of the mainframe computer. Then the PC altered our lives by making technology personal. Next, the Internet increased the velocity of change once again. I love going fast and I know how quickly things change, and even I am stunned by how many new web-based applications come out each week and how rapidly people adopt them into their daily processes. And the velocity of change is only going to increase.

Technology: For the purposes of this book, the word *technology* will refer to all software applications and all computing devices, including PCs, mobile phones, servers, networking gear, and anything having to do with the Internet—basically anything that can be used to get work done in an office environment. (I could have used the term *information technology*, but that would have added an extra word to this description—and approximately ten pages to the book!)

At the turn of the 19th century, a crude version of the telephone was our best long-distance communication device, and we were just beginning to see machines that could do calculations. Contrast that with the technology that organizations today have at their disposal: we now communicate with anyone, anywhere in the world—free of charge. Robots and machines do the majority of our manufacturing work for us. We archive data electronically and can retrieve that information from numerous office-bound or mobile devices. We have software that can perform extensive analytics on our operations and that will notify us if something is out of whack. Our workforces are spread over large geographic areas. There is no dividing line between work and nonwork; we are now accessible at all hours of the day or night. Workers from one hundred years ago would consider the tools we use today to be supernatural. And in some ways they are.

To be an effective leader from now on, you will have to understand how to guide your organization in the implementation and usage of technology, as it is a powerful facilitation tool for reaching goals most entities set. You will have to be completely familiar with the strategic uses of these tools, even though you will likely never know how to use them at a more detailed level. Saying that you do not need to understand the technology structure of your organization at a strategic level is like saying you do not really need to know anything about accounting. I suspect that even though you have a chief financial officer (CFO) on staff, you are not ignorant of basic accounting principles. Likewise, having a chief technology officer

(CTO) or VP of technology is no excuse for not understanding the basic principles of technology. Being open-minded and absorbing this valuable information will serve you well for the rest of your career.

I recently read a white paper from one of the largest accounting firms in the world, and it recommended that all businesses add a technology expert to their board. This firm observed that technology has become crucial to the success of organizations and concluded that it didn't make sense to have a board with no tech knowledge because it wouldn't be able to hold management accountable for tech innovation. I got this picture in my head of a board with fifty- and sixty-year-old members, none of whom knew an access control list from a stored procedure—but nevertheless, I was encouraged to see a large firm with influence trumpeting the need for people in the room who have technological savvy.

I'm not saying that you need to know how to write code for Microsoft Word. But you do have to know how to use Microsoft Word's basic functions. More important, you need to know the role Microsoft Word plays in your organization. You have to know whether you should pay to upgrade your staff to the latest version of Word, or why you might want to let some of your staff move to Google Docs as their word processor. Now, if you look at this example and say to yourself that you have no business being down at that level of document processing, then you may not understand the importance of the productivity that is gained or lost when your staff spends millions of collective hours using these tools. In today's high-velocity environment, your word processing system is a key component of your operations.

In fact, technology affects so many aspects of your organization that we need to discuss several different areas in which it makes the largest impact. It is my hope that this discussion will help you optimize how your organization "does" technology.

COMMUNICATION

Since the mid-eighties, improvements in communications technology have been on a quick curve up; we now have audio, visual, and text-based communication systems that include email, text messaging, instant messaging, and microblogging. The Internet has opened up an increasingly specialized set of tools for communicating in various ways that serve specific needs. For example, email allows us to instantly communicate to one or many people with the press of a button—free of charge. Text messaging allows us to communicate in short messages that can be sent and received on mobile devices while we're taking part in other activities. Microblogging services like Twitter allow us to create streams of information that can be voluntarily joined so that we can efficiently follow information providers. In addition, bandwidth speeds are improving to the degree that video communication over the Internet is now possible, free of charge and in fairly high quality. We now have a suite of communications tools that allow us to converse with virtually anyone, anywhere, anytime, for almost no money.

It is critical that you, as a leader of today, create sophisticated strategies to leverage all of these capabilities in your company. I will address social technologies later, but even outside of social media and networking, organizations are not fully leveraging the basic digital communication tools at their disposal. Many of my clients are actually having to practice reverse mentoring, in which young people teach their elders how to use these tools because the "leaders" have never invested the energy to learn how to use them.

GENERAL PRODUCTIVITY

The economies of all the developed countries are quickly moving toward knowledge-based operations and a predominantly white-collar workforce. Even manufacturers are finding that increasingly mechanized production has resulted in a higher percentage of staff dedicated to office processing. Technology helps these workers perform more efficiently with higher-quality outcomes. This includes everything from the simple task of creating a document to sophisticated electronic work flows that govern hundreds of steps formerly managed by humans. Statistically, productivity has been increasing in the United States for years, and this trend is unquestionably driven by technology. New concepts like "white collar lean," ad hoc work flow systems, and "crowdsourcing" are part of a powerful set of tools that allow us to perform tasks more quickly and efficiently.

So now on to what leaders need to understand: if you are unfamiliar with white collar lean, ad hoc work flows, and crowdsourcing, you aren't really leading. None of these concepts is brand-new. They have been talked about, written about, and put into practice by many organizations over the last few years. I will be talking about each of them later in this book. Understand for now that they are perfect examples of how technology is changing how we get things done, and how we can leverage new technology tools to be more productive.

TRANSACTION PROCESSING

After email, the first major use of the Internet was ecommerce—the ability for banks, retailers, and consumers to make transactions over the Internet. It also allowed businesses to transact with each other over the web. But even today, a decade after ecommerce became commonplace, organizations still lack sophisticated means to make transactions with their customers and vendors. Most are satisfied with simply allowing people to place an order and track the shipping, or something similarly rudimentary. This is the most basic level of a transaction, and seeing so many companies stop there begs me to explain that there is much more that technology can do. We'll cover the specifics later, but know that technology has enabled completely new methods for doing business with any person who interacts with your organization. Many companies barely scratch the surface of the value they can provide with the technologies available to them.

Leaders today must have firsthand knowledge of how their constituents can do business with them online. That includes understanding the buying experience, all the way through the delivery process and the follow-up satisfaction survey. It doesn't matter whether you are a lawyer, a doctor, a retailer, an association, or a government: you are transacting business online in some way—or should be. Sadly, most leaders might be able to tell me that they are doing transactions online, but many have no idea how the sales take place, what the experience is like for the user, or what improvements could be made. As an analogy, think about these common business scenarios:

- Do you think any leader would fail to visit a retail store that they own?

- Can you imagine a service provider who doesn't care what their office space looks like when clients come in?

- Do you think any leader would ignore a customer service rep who was rude and unhelpful?

Of course not, but when it comes to the online equivalent, leaders are often ignorant of what goes on in their companies.

DATA FLOW

Technology now allows us to control how data is created, stored, and delivered, which gives companies who leverage data well a huge advantage over those who don't. Today, we can automatically gather unbelievable amounts of data about the smallest detail, store it forever, look it back up, and deliver it to any device, anywhere in the world at a moment's notice. This massive amount of data, when analyzed effectively, can be transformed into invaluable wisdom about your operation.

What leaders need to understand is the "flow" of data. Data flows like water from place to place, and the plumbing it flows through can be sophisticated or flawed. The data can be analyzed well, or it can be analyzed incompletely. It can flow to the right people at the right times, or it can remain hidden on a hard drive, where it is virtually useless. Leaders must have a feel for data flow and see that it runs through an organization like fresh water through a faucet. They must understand what inhibits this flow from being optimum. They must understand the risks to the organization if data is lost or stolen, and know how to prevent the situations that can bring that loss about. Leaders must also have an ability to think of new types of data that can be gathered and the methods by which it can be leveraged.

This becomes a "feel" thing for a leader, more than an engineering skill. *Even without a tech background*, you can develop a sense for the underlying blockages that cripple the ability to get information to the right place, at the right time, in the best possible way. As you read the following chapters, you will gain this sense of how data flows around your organization and where the weak points and blockages are—a skill lacking in most leaders.

DATA ANALYSIS

Data, in and of itself, is pretty useless. As I mentioned previously, it is only when data is turned into wisdom that we can use it to make good decisions. Technology gives us an amazing ability to analyze the underlying data, but most leaders are happy to get a few vague reports and a dashboard of numbers, and that's as far as they get. With the applications we now have, we can program customized dashboards for each executive team member with analytics that are unique to his or her needs. We can write rules that watch the underlying data and trends and then alert us to anything that escapes the boundaries. We can produce complex ratios and express them with data visualization techniques that allow us to quickly see trends we never knew existed. These tools give us powerful insights that we have never had before now.

A powerful leader understands the importance of analyzing performance and market trends so that crucial adjustments can be made. Without a working knowledge of how sophisticated analytics can be implemented and how data can be visualized, a leader can flounder in a sea of data that shows them very little of the true story. And don't be naive and tell me that your IT department has a handle on this one! They may know technology, but they do not know what you need to see in order to make good decisions. They often barely understand what the true business drivers are; more often they are busy keeping the email server up and running or repairing a PC infected with the latest virus. Ask yourself these questions: As a leader, can you find any piece of information you think would help you run the organization? Can you create a new report and have it delivered how you want, when you want? Can you customize your real-time dashboard with new variables so you can track performance? A relevant leader of the future will be able to do these things as easily as sending an email.

I often say that technology is just a tool. But it's a tool that is almost magical in the right hands. It might be more accurate to say that technology is like a platform that supports almost every organization, and we're moving to a time when technology will be even more integrated into the operation of our businesses at every moment. It's challenging to come up with an apt metaphor to get across how important it is for *every* leader in *every* industry to understand technology as a tool in their operation. Maybe a sports analogy would be helpful.

Would it make any sense to hire a professional football coach who had no understanding of the equipment, training methods, or game plans that the team would need to know to win? I know, some of you will argue that a good motivator can have assistant coaches who know such things and get by. And I might buy that, except for the fact that the head coach is supposed to be able to judge the skills of the assistants and make the big decisions concerning the team and games. How could a head coach possibly do this without knowledge of the basics of football? He couldn't—yet we allow companies to be run by executives who have no clue about technology architecture, usage, or strategy.

If this hits home for you, fear not: the following chapters are going to give you a user-friendly crash course in leveraging technology.

CHAPTER 2

THE IMPORTANCE OF DIGITAL PLUMBING

There are three kinds of death in this world. There's heart death, there's brain death, and there's being off the network.

—Guy Almes

IF THERE IS ONLY ONE THING YOU TAKE AWAY FROM this book, I hope it is the concept that as a leader you are responsible for improving the digital plumbing of your organization. That means that at the end of the day you—not the IT department, nor the VP of IT, nor the chief information officer (CIO)—must understand, drive, and be accountable for how technology is structured in order to reach the strategic goals of the operation. As you will hear me say a few times, technology enables velocity—the speed of getting products to market, the speed of delivery, the speed of analytics, and the list goes on. Speed is our friend in almost every case. An organization's digital plumbing is what facilitates this speed, and it has become the single most important variable for success in many organizations. If you are the leader, you must take ownership of the digital plumbing—you must hold it, love it, and know it. You can outsource the construction and maintenance, but not the strategy.

So what is digital plumbing? It is the technology system that allows information to flow through your organization. If you don't know when your toilet is stopped up, your house is going to flood. In the same way, if you don't know when your digital plumbing is clogged, your organization is going to be damaged.

> **Digital Plumbing:** The technological framework that allows information to flow through your organization.

Let's talk a little more about technology in general before we move on to practical advice on how a civilian leader—one who doesn't have extensive technological knowledge—can learn the digital plumbing model and use it to make dramatic improvements.

As mentioned before, technology is just a tool. But it is "just a tool" in the same way that a piano is "just an instrument." There is a huge difference between one of my kids banging out a few notes on the piano and a concert pianist performing a sonata by Beethoven. Same tool, different person— very different result. Assembling the digital plumbing within an organization is exactly the same. There is a ton of variability in how any specific IT team will construct and operate digital plumbing even when they are given the exact same mission. I have assessed the technology infrastructure of many different organizations in the last twenty-five years—all sizes and fla-vors—and I have seen wonderful masterpieces. But mostly I see broken, leaky plumbing systems in which data does not flow and users are crippled in functionality. For leaders to simply outsource digital strategy to contractors or in-house IT people is just irresponsible. Yes, I understand that this is a whole new skill set for most leaders, but try learning something new. It won't kill you.

When I get hired to do technology assessments for organizations, I always do a survey with about a dozen of the company's employees. I ask the same questions each time, including how well, on a scale of one to ten, they think the organization utilizes technology. The average number I get is around 3.5. Then I ask how dependent on technology the organization is to do what they do. (A score of 1 would mean they could run the whole business on a legal pad; 10 means they could not even open the doors and work without technology.) The average number I get nowadays for this question is 9.2. Think about that for a second. It means that most people can see huge areas for improvement in the organization's digital plumbing, and at the same time feel the organization could barely function, if at all, without the digital plumbing.

Given these numbers, it's amazing that some leaders tell me they can't see how making technological improvements would help their businesses that much. This shows me that they are out of touch with the real operation of the business, and that they lack imagination for how improvements could be applied. Other leaders just copy other organizations. They wait until they see a good idea from a competitor and then rush to buy the same capabilities. The world of today's technology provides a huge and wonderful palette of tools, and they are only going to get better and more plentiful; what a tragedy that so many leaders cannot see how improvements to the digital plumbing of their organizations could help them win.

Since keeping pace with the competition is vital, learning how to leverage technology and generate the velocity you need to win is the only sane thing to do. The good news is that companies can change: I have watched a number of organizations move from IT chaos to operating as a well-oiled machine in a matter of months. This was accomplished by creating a new vision in employees' heads as to what needed to be accomplished and by adding new processes for managing and utilizing technology. A logical place to start changing the way your organization uses technology is to reframe the role technology plays in the first place.

The first thing you have to do is consider the big picture. Because technology is now all around us, it has created a "can't see the forest for the trees" situation. We interact with technology from the time we wake up until the time we go to bed. Different people in our organizations use technology for different functions: accounting, data storage, communication, and the list grows every day. The problem is that most leaders understand the individual uses of technology (the trees) but don't grasp the overarching technological strategy (the forest).

So what is the big-picture view of technology in an organization? Envision it as a collection of interconnected mechanical devices topped with a layer of software that delivers specific applications. You can also think of it as a sophisticated matrix with thousands of ways to move data where we want, when we want. This sophisticated matrix must allow data to flow and must provide security so that only certain people can tap that flow. It must also provide backup capabilities so the data is never lost, and it needs to provide analytical tools to leaders so that they can glean important facts and trends from the data flow.

Why is it important for you to have this big-picture view? Most fundamentally, it's important because we want to get the most out of the tools available to us. Based on my broad experience working with technology and advising clients, it is clear to me that most organizations that invest serious money in software rarely use even 50 percent of the capabilities they purchased. There are many reasons for this. Sometimes leaders simply do not believe they need to use any more than what they're using to be successful. In some cases, they simply do not understand what the technology could be doing for them. In other cases, they understand what the software does generically but cannot make the leap to coming up with specific ways to use the tools. It is disheartening when I see clients who have committed the funds to buy good plumbing and then fail to fully implement it or use all its capabilities. It's like having a hot tub in your house and never using it because you didn't take the time to fill it up or learn how to turn it on.

For a good example of how we don't always take full advantage of technology, look at the wonderful and inexpensive reporting tools we have for expressing and analyzing the information in our databases. For most people, a "report" means a collection of paper that is printed out and put on their desk. But to a computer, a "report" could be delivered on paper, in an email, through a text message, on a web-based dashboard, or in a webpage. These reports can be pulled when needed, delivered based on a condition being met, or sent in any increment you like: daily, weekly, monthly—you get the idea. Would you rather have a flat collection of information pulled straight out of a database, or a report with many layers of data built into it? The difference is between a sales report that simply states a salesperson's sales for the month of May and a report that provides the same information, plus the analytics to show what percentage of their monthly goal they achieved and their numbers compared to last year. (The second option is much more valuable, obviously.)

Most people who aren't familiar with technology have little ability to imagine how these reporting tools can be used; hence they do not even know to ask for reports that could easily be produced. They ignore fantastic capabilities that end up going to waste. I cannot tell you how many times I have heard IT people express frustration that the leaders in an organization do not tell them what information they need to run the business or what reports would be helpful. Then, when I gather a list of reports the executives could use by asking lots of questions and helping them understand what is possible, the execs are mad at the IT people for not asking the same questions. When you consider that there are organizations with hundreds of employees and millions of dollars' worth of software not being used to maximum potential, the waste of resources is staggering—but not surprising when you understand that there are many knowledge gaps between executives and IT people.

Information must be able to flow through your digital plumbing in an effortless and dependable way; when it does not, there will be a lot of frustration and angst. Every time I turn on the faucet at my house, water comes out. This is the standard we want from the technology used in our organizations, but we rarely receive it. *With dysfunctional plumbing, the cost in productivity and good decision making is a hidden burden that organizations rarely take the time to understand.* Improvements can be made, but it takes a conscious effort to examine all the elements of your technological infrastructure and fight for the necessary renovations.

CHAPTER 3

THE STRUCTURE OF YOUR DIGITAL PLUMBING

I can't understand it. I can't even understand the people who can understand it.

—QUEEN JULIANA OF THE NETHERLANDS

NOW THAT YOU UNDERSTAND THE CONCEPT OF digital plumbing—the system that helps move valuable information to the right place at the right time, and archives the information for as long as you need it—let's take a look at its structure. In order to get a good picture in your head, think of digital plumbing as a series of layers, each built on the one below it. Study the graphic on the next page for a moment before we move on to a discussion of each of the layers.

Business Intelligence Layer
Dashboarding systems, data analysis software, data mining applications

Application Layer
PC-based software, web-based software, server-distributed software

Database Layer
Data archives, image stores, transactional data stores

Hardware Layer
Computers, servers, storage drives, backup systems, handhelds

Communication Layer
Bandwidth, routers, firewalls, switches, network infrastructure

Let's start at the foundation of digital plumbing, the **Communication Layer**. It connects computers so they can communicate with each other—through bandwidth, routers, firewalls, switches, and the network infrastructure. These are the pipes that connect systems to the Internet or a data center. To continue our plumbing analogy, you could imagine this as the layer that connects your house to the sewer system outside. It is normally underneath the ground where you can't see it, but if it ever backs up, you know about it quickly. It is the same with the communication layer: you know right away when it's down because you lose your Internet connectivity and email in an instant.

The next layer is the **Hardware Layer**—computers, servers, storage drives, backup systems, and handhelds. This collection of gear provides the physical means of managing and interacting with information, and it allows us to store massive amounts of data. These are the vehicles we use to run our applications, and they provide systems that filter certain users from accessing our technology and prevent certain information from entering our system. We carry the hardware that comprises this layer on our belts and in our bags, and it sits on our desktops. Almost every device in the hardware layer is now connected to the Internet in some way, whether through a cell phone tower or a data center. Of all the layers described here, leaders are probably most familiar with the hardware layer.

The **Database Layer** stores the raw data we create in data archives, image stores, and transactional data stores. This layer is critical because it also gives us tools to create new views and reports by mining the data we have gathered. One exercise I go through with some clients is having the IT department draw a map of all the databases in the organization and then reviewing the map with the management team. The managers are always amazed at the complexity, the size, and the chaos of the data storage layer. Even civilians with no technological savvy can look at a database map and see the issues that have been created by installing many different applications on top of the hardware layer—all with their own databases, which many times do not "play nice" with each other. This layer is becoming more and more important as organizations struggle to improve the way they gather, store, and manipulate data. When leaders do not understand the database layer, they cannot get the dashboards, reports, and analytics they need to make wise decisions. And the larger an organization gets, the harder it is to manage the database layer. Leaders normally end up with huge amounts of data on separate servers, and have little means to normalize the data across databases. Normalizing data includes tasks like creating consistent field names so that data can be more reliably searched. It also means de-duplicating records so you do not have multiple versions of the same underlying information. Organizations end up spending millions of dollars on data warehouses and specialized data manipulation applications to solve the data normalization issues that plague them.

Thankfully, smart people have been developing standards in the computer industry that allow much of this plumbing to be connected easily. These visionaries are rarely publicized and get little credit for the important role they play, and they are often simply giving away their ideas for the greater good. This sharing of ideas was not always present in the past, though; back in the day, many companies tried to create their own proprietary standards to force buyers to adopt a specific company's products. Thankfully, thousands of IT buyers and builders over the past two decades have driven vendors to adhere to universal programming and data-handling standards—but the battle continues today.

The next level up is the **Application Layer**, with its PC-based software, web-based software, and server-distributed software. Applications are the visual interfaces we see on our screens every day. They allow everyday users to do something meaningful with the information stored in the layers below. There are small applications, like widgets and gadgets, and there are huge applications that can serve as the all-encompassing technological environment for large businesses (products from SAP and Oracle, for example). Applications provide the interfaces that sit on top of our databases and manage the flow of information in and out of our storage systems. They also help us communicate with each other, calculate "what if" scenarios, and suggest solutions to problems. There are millions of applications in the world today, but just fifty years ago there were probably only a few hundred. Even more mind-boggling is that every week hundreds more applications appear on the market. We are swimming in a flood of new applications, and no single human can possibly keep up with what is new on the market. Organizations must continue to learn about new applications, however, and adopt those that will prove most helpful. I often ask my clients what kind of research and development (R&D) process they use to discover new applications that could help them. (When I use the term R&D, I am referring to the practice of having people who are responsible for studying what new software and capabilities are coming on the market.) Not surprisingly, almost no one has a formal process for doing this. In a world where the trend is toward free and cloud-based software solutions that are inexpensive, it makes sense to have an established R&D process to discover new tools to manage data flow. When this process does not exist in an organization, it can quickly fall behind in its knowledge of which tools could best help its digital plumbing be more productive.

Cloud Computing: The practice of providers like Amazon, IBM, Google, or Microsoft renting out servers and storage in giant data centers on an as-needed basis. This practice replaces the need for an organization to have an internal data center.

The top layer is the **Business Intelligence Layer**, including dashboarding systems, data analysis software, and data mining applications. This is the least understood and least utilized layer at the moment but it is quickly gaining popularity. Large organizations have been paying attention to this layer for years now, and small and medium-sized operations are beginning to get the picture too. In this layer, we build tools to analyze our data and information and represent it in real-time dashboards that give people a useful glimpse into the performance of an organization. We also use this layer to build rules and controls that automatically take actions based on conditions set by employees. This layer will one day be highly customized for every organization and will be one of the most expensive assets to purchase in an acquisition. The institutional memory and knowledge now held in the minds of humans will be transferred to this layer—with huge impact on the scalability of organizations. I will dedicate more time to a further exploration of this layer later in the book.

All of these layers have a sophisticated set of pipes that connect both internally, between servers and organizational locations, and externally, to vendors, customers, service providers, and others. Every organization depends on its digital plumbing for data and communications, and no organization can survive without it. Yet, even though many leaders understand this concept, very few focus on building world-class plumbing. They simply outsource the work of strategic technology design to contractors and IT staff and hope that the chaos gets sorted out over the years. As a leader, you cannot afford to do this anymore. You must have a strong working knowledge of your digital plumbing before you can begin to optimize its performance, and the performance of your organization.

Note that the size of the organization is not a factor in the need for good digital plumbing. Even if you are the leader of a five-person firm, you are almost certainly performing tasks electronically, and therefore need to have a strong technological framework for your employees, vendors, and customers.

CHAPTER 4

FINDING THE LEAKS IN YOUR DIGITAL PLUMBING

We are too busy mopping the floor to turn off the faucet.
—ANONYMOUS

FOR SOME OF YOU, IT MIGHT BE EASIER TO LEARN HOW to improve your digital plumbing by looking at the ways organizations are struggling today; it is difficult to fix something if you cannot identify why it is broken. Many companies are far enough along in technology development that their leaders have all the right equipment and skills available to them, so the only factors holding them back are lack of perspective and the will to change. To begin with, let's take a look at how the various areas of your organization could be improved by paying attention to your digital plumbing.

As you read through this list of dysfunctions, see if any of them strike home. If so, write them down. You will need them later when I show you how to put together your technology strategy (see chapter 6). It is easy to identify which parts of your digital plumbing are broken; just look at the list of things that annoy you, the things you ask an IT person about—why something cannot be done, or whether it could be done faster—that always result in you getting double-talk or the runaround.

> **Digital Plumbing Leaks**: Technological weaknesses in your company that, although fixable, result in losses of some kind.

COMMUNICATION LEAKS

One of the most dramatic impacts technology has had on the world over the last decade is the explosion of tools that allow us to communicate with each other through a computer or mobile device. Think about how quickly we have become dependent on applications like email, text messaging, Twitter, Skype, and blogging to communicate with each other. And that's without considering the impact of cell phones! In the space of fifteen years, we have gone from virtually no practical ability for affordable, real-time conversations across the globe to instant communication with almost any corner of the world at little or no cost. We went from limited, tightly controlled systems of publishing one's thoughts to the world to millions of people sharing their ideas through online profiles and blogs. It is no overstatement to say that people will one day look back and say that the dividing line in communication we have just crossed was one of the most important for the human race. We may want to change era designations to "before digital plumbing" (BDP) and "after digital plumbing" (ADP).

But even ADP, most organizations struggle with getting team members to communicate effectively. All of the communications tools just mentioned have helped facilitate a huge improvement in our ability to communicate over long distances and build new relationships. At the same time, they have also altered the dynamics of face-to-face communication in ways that are not always healthy. We now see managers trying to use inappropriate tools to counsel with employees, kids who use texting to say what they would never say in person, and spammers who do their best to hijack every communication path they can in order to sell products most of us would never buy.

Organizations rarely study how the institution communicates internally and externally with an eye toward what can be done to improve the general flow. We now have a grab bag of technologies that facilitate conversations, and some of these might be monitored—usually by the IT department—while others may not be. We may legislate when and how certain communication technologies can be used, but for the most part, people are left on their own to choose the tools they feel comfortable with. What productivity gains could come from closely studying how communication gets done in a specific organization, with the goal of creating an optimal institutional paradigm? Would the bottom line be improved 2 percent, or would it be more like 20 percent? My guess is that the results would be closer to the 20 percent end of the spectrum. How can we see this improvement by fixing the communication leaks in our digital plumbing?

Let's look at email as an example. How much time is wasted by people who copy half the company on unnecessary emails? How many people send long, involved emails where a phone call would do? How about the people who send three-word emails that don't answer the question and force a phone call for clarification? Email can waste a significant block of time, but many organizations make no attempt to set institutional norms for handling it. They just accept the waste.

We are certainly moving toward a world where we have a single communications server that will provide consistency in how all forms of technology-aided communication takes place. In the meantime, organizations could benefit in a huge way by taking a step back and designing a plan for how they would like all communication to happen and then working to embed this plan into the corporate culture and the digital plumbing. The goals of the plan are to improve the speed and quality of information sharing, to monitor what needs to be tracked, and to automate as many conversations as possible. All of this will drive down the cost of your organization's communications.

In addition to improving the use of existing communication tools, leaders need to be aware of the new communication tools that are impacting our organizations. We have wonderful new services that not everyone is aware of or comfortable using, which has led to a younger crowd that communicates through a very broad set of tools while the older crowd lags behind. Think about it: if you have access to the Internet, you can use a telephone, email, text message, instant messaging, microblogging (Twitter), videoconferencing (Skype), or blogging as methods of communicating both internally and externally. Depending on their proficiency with these tools, people develop personal preferences for which ones they use—and this creates unnecessary chaos in how organizations communicate.

Another important goal is to improve the *quality* of communication. Different kinds of information need to be communicated in different ways. There are certain types of communication that need to be done by humans, and there are other types best handled by technology. Knowing the difference is important. For example, it makes no sense to have an employee call and report that a shipment has been sent if an automated email can serve that purpose. On the other hand, it also makes no sense to have an automated customer support system if your customers hate it. Once again, you see that technology has given us wonderful tools, and we are struggling to find the most productive ways to use them.

My guess is that the pendulum has swung too far toward the use of electronic tools because they are new and viewed as highly efficient. We will soon learn, however, that other traditional forms of communicating (such as in person and over the telephone) are more effective for certain things, and we will trend back in that direction in some cases. The bottom line is that your digital plumbing needs to be able to support a rich collection of communication tools that are right for your organization. And as a leader, you need to determine the communication leaks in your digital plumbing and drive the improvement of the system.

Visit www.velocitymanifesto.com for more information on social technologies and virtual team building, both of which are also covered in more detail in chapters 16–19.

BACK-OFFICE LEAKS

One of the most target-rich ways to gain a return on investment with technology implementation is to apply it to the back office. For our purposes, we can define the "back office" as the support services customers never see. For instance, accounting, product development, legal, operations, administration, and human resources. In most organizations, large and small, a heavy percentage of the overhead costs are tied up in back-office personnel costs, and an improvement of digital plumbing in the back office will have a huge impact on profitability.

IT Back-Office Costs: Costs associated with an organization's IT and infrastructure systems, as distinguished from the costs of the company's public-facing products and services.

Stopping Sales Leaks

Let's start by looking at how customer relationship management (CRM) technology impacts salespeople as an example of how digital plumbing can be improved to drive better bottom-line results. CRM allows salespeople to be more effective by providing structure to the sales process and the archives of information about customers and prospects. It also provides quota information so salespeople are crystal clear on what is expected of them and know how they are performing at all times. Electronic marketing processes help develop leads at a low cost so that the CRM systems are continually fed with prospects that can then be sold. Real-time sales information can be pushed directly to the salesperson's handheld device so she knows immediately if she has a customer service issue. CRM has redefined how sales departments are run because it provides a much more timely and predictable method for gathering sales pipeline data. With CRM, salespeople also have much more visibility into what is going on at their company through software that allows them to keep track of manufacturing, distribution, and billing without calling someone back at the office for research. The added communication options of email, RSS feeds (real simple syndication)—a service that allows a user to connect to a stream of content so that it is always pushed to them in an RSS reader or to their email client—and websites have given salespeople additional tools for communicating with customers in the ways that are most efficient for them. As we'll continue to see throughout this book, sophisticated digital plumbing can have a huge impact on top-line revenue when constructed correctly. I will talk in more detail later about other ways that technology is changing the sales process.

Stopping Management Leaks

For managers and executives, good plumbing can provide an improved quality of information and analytics so that they can make better decisions. There is a still a huge gap between the raw data we generate and the useful knowledge that should be available when it is needed. In many cases, huge improvements can be made simply by optimizing the use of the reporting system that taps into the data layer. I find it tragic that some organizations have invested huge amounts of money in building databases and buying sophisticated reporting systems but then fail to take the last step of training people on the use of those reporting tools. Many times I see IT departments hold the reporting system hostage so that users must come to IT for a report. This makes no sense, because we now have reporting systems that are one level removed from the data so that users can view information without danger of damaging the underlying database. By training users in these tools and giving them free rein to mine the data in whatever way provides value to them, progress is made.

Beyond simple reporting is the ability to electronically monitor the performance and time usage of certain types of employees. Putting this capability to use allows managers to better understand productivity gains and losses. They can build rules into applications to help them control any issue that could injure the bottom line. For example, managers can develop controls that require their approval on any orders sold at less than a desired profit margin. This helps stop unethical employees from abusing access to sales systems and keeps profits from dripping out of the bottom line. In most businesses, there are dozens of such rules and controls that could easily be programmed into software systems that would help prevent undesirable events. The missing link is managers' and executives' understanding of how easy it is for programmers to write small pieces of code to accomplish whatever needs to be done to provide more information or stop up a leak.

Stopping Operational Leaks

Probably the largest gains in using a watertight digital plumbing system to prevent back-office leakage are to be made in operations. A huge amount of payroll is spent on paying people to do tasks by hand or paying other people to monitor them with human eyes. Examples of these tasks include the following:

Multiple or manual inputs of data. Messy methods for inputting data can result in high costs and low-quality results. If any of your employees are filling out forms for customers or reentering data from computer-generated forms, you should automate the process. For example, insurance companies send out explanation of benefits (EOB) documents to doctors' offices with payments. Most doctors' offices then have to rekey them into their accounting systems by hand. This is a labor-intensive process that creates potential for the mis-keying of information. The doctors' offices should be looking for a way to connect their accounting system to the insurance companies directly, or to have their banks lockbox the payments and EOBs and send over both electronically. This digital plumbing eliminates the need for hand keying.

Working with software applications that do not communicate with each other. If your employees need to key the same information into two or more different systems that do not communicate with each other, you have a leak in your digital plumbing. A good example of this might be a bank that has purchased four separate software applications to run its general ledger system, its lending operations, its savings accounts, and its investment accounts. Imagine that a customer walks in and wants to open an account, borrow money for a car, buy a house, and open an IRA. The customer may have to visit a few different desks in the bank, and while doing that will have to fill out several sets of paperwork with the same information. Then the bank employees that work in each of those departments will all have to key in the same information by hand. Six months down the road, the bank manager will not be able to type in the customer's name and access one report that lists all the business the customer does with the bank. Nor can the marketing department do any sophisticated customer research to know how they could best up-sell certain customers.

In a world with the best possible plumbing, new customers would go to one screen that would set them up to do online banking. While entering their information on this screen, they would choose from a list of products they need from the bank. As they do that, only the information needed for each of those products would be put on the screen to fill out. Then they would provide their e-signature and complete the transactions. In fact, this could be done from home before they ever step foot in the bank or meet their representative—who, by the way, would help them with any product they buy from the bank by being a single point of contact and reaching out to departmental experts when needed. The interesting thing about this example is that by improving the underlying digital plumbing, we also have the opportunity to change the way the bank works with customers.

This is not a new concept. We have been implementing electronic data interchange (EDI) standards for decades to facilitate the exchange of electronic documents like purchase orders, invoices, and payments. Other examples of EDI implementation include insurance companies filing claims electronically, brokers facilitating stock sales online, consumer products companies receiving orders from large retailers, and just about any form that is filled out from a website rather than by hand. In order for many white-collar businesses to prosper in the next decade, they will have to learn the skill of ruthlessly examining what roles humans are playing and try to mechanize them. Our first step in this direction has been the massive attempt to outsource these tasks to countries with a lower salary base. In many cases, we are using digital plumbing to facilitate the integration of foreign workers into our work flows. The next step is to write software that is intelligent enough to eliminate the repetitive tasks completely.

Let's go back to the example of electronic claims processing at an insurance company. Technology has advanced enough that the claims can now be transmitted in electronic form and then passed through the insurance provider without anyone having to key them in by hand. This is a huge time-saver, and therefore a big benefit to the bottom line. The next step is to write intelligent logic that fully mimics what the human processor would look for in order to process the claim. Then allow 95 percent of the claims to just run through the logic engine and go directly to payment (which is called automated adjudication). Kick out the 5 percent that the computer cannot use logic to sort out and a human can apply their knowledge to deal with them. Every business must reexamine its staff regularly and determine what tasks are unnecessarily being done by hand.

For more information on cutting back-office costs, see the sections on white collar lean (in chapter 12) and crowdsourcing (in chapter 11), and visit www.velocitymanifesto.com.

ANALYTICS AND DECISION-MAKING LEAKS

Once digital plumbing allows information to flow smoothly and seamlessly through an organization, the information must be harnessed so that good decisions can be made. I cannot say this strongly enough: as far along as most businesses might think they are, there is still a big gap between the useful information that *can* be reported and the useful information that *is* reported. The managers and executives who hold the knowledge of what would be helpful for making good decisions often do not know much about what the software they already own could do for them. In general, we often use a mere fraction of the capabilities provided by the software we buy, and we also have little imagination for tools like reporting services.

The result is that we waste some of the most valuable tools we have assembled as part of our digital plumbing. I've heard many stories about companies that learned how to apply business software and were shocked to find negative anomalies when they ran the first set of reports. Imagine leading a company where sales were driven by catalog mailings. In order to more efficiently target particular regions, you decide to produce five different catalogs with different price ranges and types of goods instead of the one major catalog you have produced for two decades. You start analyzing sales patterns by zip code and designate which products will sell best in each area; then you allow your software to assemble hundreds of variations of the catalog to maximize order flow from each zip code. This all starts to sound pretty sophisticated, but with the right software it can be done without too much of a time investment. The shame of this is that you have always had the underlying sales data and have always known that different zip codes bought slightly different types of goods—you just didn't apply software to this useful data so you could automate the creation of custom catalogs for each sector.

Another example is real-time online dashboards. Every manager and executive should guide their team based on real-time results. In order to do this, each person must create a web-based page of information that shows exactly what is going on in the areas they run. Managers should have the ability to drill down this information so that when an anomaly presents itself, they can quickly click through to the underlying data in order to diagnose what is going on. This is a must for *every* manager and leader.

Dashboard: A real-time, web-based set of indicators that gives the user relevant information about an organization, giving that user critical data for making sound decisions.

For a good example of the usefulness of dashboards, let's look at a regional manager of a group of Subway franchise stores. He should be able to call up a dashboard that shows a list of stores along with each one's sales for the day. He should be able to see the day's sales in comparison with the previous day, or any day a week, month, or year in the past. He should also be able to drill down to see the average ticket price and the number of customers for the day, as well as the day's profit margin, with comparisons to historical data, and maybe even analytics like revenue per employee or ballistic information on product sales (the rate at which sales are growing or shrinking). If our regional Subway manager is on a trip to Florida for a conference, he should be able to walk up to any computer with a web browser, call up this dashboard, and see how things are going. Moreover, he should be able to call this dashboard up on his handheld device from anywhere, at any time. If this strikes you as too complicated to build, you should realize that it's being done in many organizations today. The bottom line is that if you are a manager or leader without access to a real-time dashboard of information, you are underprepared to run a team or an organization. This kind of dashboard can usually be created by drawing it up on a piece of paper and explaining to your IT department what you'd like to see. Here are a couple of specific problems that plague the analytic operation of your digital plumbing:

Databases that are not normalized, or that cannot even share data. Each application has its own separate database that will not pass data to other applications' databases. The most efficient setup is to have a central database that contains customer information and also to have separate smaller databases that hold specific product information linked to the central database. It is critical to understand that each time we buy a new software application, we are also usually buying a database that sits beneath it. Over the last twenty years, many organizations have created a huge mishmash of databases that do not interrelate. This is a prime example of dysfunctional plumbing.

Missing software applications. I am always surprised when I see basic applications missing from my clients' operations. I have seen very large organizations managing their sales force from a spreadsheet instead of a CRM system. I have seen companies with a total lack of reporting systems, so that their managers have no way to judge the performance of the people on their teams. I often see websites built in straight HTML so that no one can edit them but the IT people. Leaders tolerate these leaks in their digital plumbing because they have no picture of their overall plumbing system, or because they are so busy trying to deal with what they have that they do not realize what they do not have.

If you want to be an effective leader, you have to develop the ability to look at the entirety of the technology structure in your organization and have a sense for what is missing. I wish I could give you a list of twelve things to look for, but the specific plumbing for any operation is highly customized. What a bank needs is very different from what a hospital or a manufacturer needs. If you are the leader of a bank, you need to learn what the digital plumbing diagrams need to look like at a bank and then, when you move to a new job, you will be able to quickly look over the system in place and determine whether something critical is missing. Leaders in an organization have to make decisions. I've observed that many of them make these decisions by gut feeling alone way too often. The data to help them make rock-solid, fact-based decisions exists; they simply lack the ability to get the IT people to provide what they need, or they lack the creativity to imagine what helpful analytics they could ask for.

We'll cover more information on creating better visibility into the operating performance of your organization in chapter 8, which covers the concept of business intelligence. To learn more about this concept, you can also visit www.velocitymanifesto.com.

BUSINESS CONTINUITY LEAKS

As we have become more dependent on electronic data and records, the need to protect this information has become critical. A very important piece of digital plumbing is a foolproof backup system. Most organizations are conscious of this fact and have taken action to make sure that data is not lost. The next step is to make sure that we can actually use our electronic tools in a disaster situation. With the spate of man-made and weather-driven disasters that have occurred in the last five years, companies have learned that simply backing up data is not enough; you also have to be able to restore the functionality of the systems that allow people to do their jobs.

In order to protect your company from going dark for days—or even weeks—you must determine how long your organization can be down before the situation impacts your customer or client relationships. For some businesses, like banks, it is a matter of hours. In a disaster, people need to get cash quickly, and if the bank is closed they get very upset. It is the same with hospitals. Organizations that cannot have any downtime have to build their plumbing on redundant data centers in different cities or states because they simply cannot afford the time it would take to rebuild a data center in order to do business, and they often cannot operate or handle transactions by hand. (On the other hand, a broccoli distribution company might be OK not taking orders or shipping product for a few days.)

If you want to undertake an interesting exercise, ask your technology people to explain your backup procedures, and then ask them how long it would take them to reload all of your software applications if your servers were destroyed. Many organizations are very careful about backing up their data but don't really spend much energy backing up their applications. The problem, of course, is that they can retrieve their data but won't be able to use it until they have retrieved their applications. This creates a huge problem that most leaders are not prepared for. After a while, many of us come to believe that the plumbing underneath our homes will always work, just because it always has. Our attitude toward digital plumbing is getting to be the same way: as we get better at keeping things up and running, people come to depend on computer systems to do everything in the organization. Then, when the power goes off or the data center is destroyed, the organization is in immediate disarray. It may have been years since employees were in a situation where they had to do without their software applications.

The way to plug this leak in your digital plumbing is to sit down with your IT people and put some boundaries around what would be acceptable in a disaster. Then invest in what needs to be done in order to ensure that your system can handle those demands. Hold a meeting at least once a year in order to review your disaster plan. It is flat-out irresponsible for a leader today to blow off this task. Most middle managers are way too busy to care about things like disaster recovery, so it will only be put in place when the leaders dictate it. Don't make the mistake of putting in world-class plumbing and not taking the time to protect it.

Visit www.velocitymanifesto.com to download a document that will help you establish backup and disaster-recovery procedures.

SECURITY LEAKS

Finally, you need to consider the security of your digital plumbing. One way that digital plumbing does differ from traditional plumbing is in the need for information security. We normally do not care who drinks the water at our business. But we do care who has access to information about our customers. I hope I do not need to explain the negative impact of having your data breached. Not only is it embarrassing; it also opens you up to lawsuits that can be financially damaging.

Please understand this fact: if you are a leader who is not very computer savvy, you would be shocked to learn how exposed your operation really is. Even some of the largest companies in the world have been penetrated. If you are part of a smaller company, you are likely even more exposed, because small businesses typically lack the in-house expertise necessary to help protect their computer systems. I have a client that is a medium-sized company, and they built their core business system from scratch. One day a short time ago, one of their franchisees noticed that by changing one number in a website URL, he could get the employment records of a random employee. The CEO was mortified by this weakness, especially because it was brought to light by a franchisee. This is a perfect example of how a leader can get in trouble by not being familiar with digital plumbing. On the bright side, this CEO moved very quickly to close the holes on the software once the danger was understood.

Today, it is relatively easy to fix security leaks; we have been dealing with computer security long enough that we have the knowledge and the tools to patch problems quickly most of the time. The dysfunction here is that normally only the very computer literate understand the issues involved in building up security. Every day across the world, IT people and executives meet to set higher budgets for improved security. In almost every case, the executives are frustrated because they don't really understand what they are being asked to invest in. The IT people get frustrated because they clearly hear that their leaders want the plumbing to be bulletproof, but the leaders are unwilling to invest enough money to get the job done. And when the plumbing gets penetrated, the IT people are blamed. The solution is for executives to learn more about their digital plumbing and for IT people to spend more time sharing information with executives. Improving security can be complicated, but understanding the improvements on a conceptual level is not hard.

It is worth spending some time with someone who is knowledgeable about security to make sure you are up to speed on the basic vocabulary and concepts. Then have your technology people review with you what they have done to protect your digital plumbing. Ask questions like "What are the three most dangerous weaknesses we have?" and "What are the three things we need to improve this year in order to be more secure?" Repeat this process every year.

You can also visit www.velocitymanifesto.com to download a document that will help you establish effective security policies.

In every organization there is a gap between what the digital plumbing currently is and what it should be. That gap can always be closed, given enough time. The tragedy is to not understand the gap. Only by working with your technical people and asking the right questions will you be able to wrap your mind around the nature of this rift and make progress to close it. Unlike the plumbing in a home, digital plumbing is never complete. There will always be new, more effective tools that can be applied, and there will always be new needs to be met. The portion of the budget invested in technology will fluctuate over the years, but it is dangerous to ever believe you're done spending money on your organization's digital plumbing. This leads to the dependence on a specific aspect of the plumbing for years beyond its useful life, and it will often slowly become inefficient, or even dangerous, if left in place too long.

Hopefully you have been building a mental picture of what your digital plumbing looks like and have identified where some major improvements could be made. Being able to construct effective digital plumbing is a powerful skill, and I want you to have it in your toolbox. Remember that you cannot simply outsource the entire design of your technology infrastructure to people that report to you and think it will be perfectly aligned with the company's overall strategy. Only you, the leader, can pull these two areas together. The next trick is to learn how to craft a sophisticated technology strategy going forward.

CHAPTER 5

MAPPING YOUR DIGITAL PLUMBING

When people are free to do as they please,
they usually imitate each other.

—Eric Hoffer

IN ORDER TO MAXIMIZE THE PRODUCTIVITY OF ALL
the layers of digital plumbing, every organization needs to
have a separate technology strategy as part of the overall organizational strategy. I have observed this: companies that do not
have a *written* technology strategy specifying what improvements will be made—as well as when they will be made, and by
whom—often exist in a state of chaos. In addition, without a
written plan, there is no good way to measure the success of the
IT department or even to set goals for how it might improve
the company's digital plumbing. What ends up happening
is that we measure the success of our technology employees
based on the amount of time they keep the digital plumbing
up and running. This is important, but it fails to give them the

inspiration that will make them want to improve the plumbing in ways that support the organization's business goals.

Let me share with you a five-step process I use with clients to help them develop a technology strategy that takes full advantage of their digital plumbing.

STEPS FOR BUILDING A TECHNOLOGY STRATEGY

When setting out to create a written, agreed-upon technology strategy that involves both IT people and the operational side of the organization, it is good to follow a model that can be repeated each year. A technology strategy is not a one-time exercise. The underlying tools in the tech world change and improve so quickly it is now a requirement to update the strategy every twelve months. I have reduced this process to five steps that can be completed by an organization of any size. The larger the entity, the more complicated each step might be to complete, yet this formula works for everyone. I have taken many clients through this and promise you it solves a slew of problems that organizations with misspent technology dollars face today.

STEP ONE: Create a "Today" Map That Shows the Current State of Your Digital Plumbing

First, draw a map that graphically represents each of the layers in your digital plumbing and describes how the elements in each layer are organized. Be sure to include each of the five layers—Communication, Hardware, Database, Application, and Business Intelligence—discussed in chapter 3. These maps should be color-coded so that, for example, dysfunctional areas are in red and areas that work are in blue. These maps serve as a visual representation of how the plumbing is organized today.

One way to create this map is to have the IT department draw a diagram that shows all the databases in the company and what data they hold. Then have them draw lines between the ones that communicate with each other. With this map laid out in front of you, start looking for places where data is being entered by hand and also where it is duplicated across databases. Start asking yourself questions: Could we use existing data to populate other databases to ensure that everything is consistent? Can we share data between databases rather than duplicating it? Can we collapse databases together into larger databases that will allow us to improve quality and services (and maybe cut IT costs)?

Then start asking questions around reporting: What can't we do because certain data exists in a separate database? What reporting could be improved if we collapsed or connected two different databases? This type of mapping and questioning is easy to do, but is rarely asked for. It gives a quick picture to a nontechnical person of both how broken plumbing is impacting the organization and what improvements must be made. Remember, it got this complicated because you slowly brought many different software applications into your business without giving enough thought to how all the pieces would fit together. Now is the time to reevaluate.

I recently had a conversation with a person in the IT department of the Internal Revenue Service. I was told that they have over sixty applications they support and that these have been written with every programming language known to man. The reality is, no matter how good their intentions, they are dealing with a complicated piece of digital plumbing that has been built over many years, by many different departments, and they will struggle with refining their plumbing for years.

Next, you'll need to map your data storage, and there are two different issues to examine. The first is making sure that data is backed up in ways that ensure it is never lost. The key word here is *never*. The business world is rife with stories about organizations that thought they were safe because they did backups on tape drives, only to find that when they went back to the tapes to find lost data, the tapes were bad. Even worse is losing a day's worth of data because the backups were not made before the daily transaction register was wiped out. This gets to be a serious issue when you run thousands of transactions in a day and have no way to rebuild them. Executives need to sit down with IT people and have them draw out the entire backup strategy and show in pictures how all data is protected. It is a shame when data is lost simply because leaders do not invest any time in holding technology staff accountable for good backup plumbing.

Another very important thing to be aware of as you examine your map is the need to standardize your technological tools. Just as the traditional plumbing systems have standards— the size of pipes, the construction of toilets and sinks—digital plumbing also has standards. The difference is that physical pipes have been around a lot longer than technology has, so there has been more time to agree on the standards. Because the world of technology is still so new, and because we have multiple vendors that want to set their own proprietary standards for financial reasons, we often have disagreements and confusion about which standards to use. For example, people have fought over whether they should use Linux or Microsoft operating systems, or whether they should program in the

Visual Basic or C languages. It's frustrating to be in a situation where IT people want a decision on which standard to use when the executive has no idea which way to go. There is no good advice on how to resolve debates on standardization other than to listen carefully to the motives behind each debate and make sure the decisions are being made based on which technology standard is best for the organization in the long term and not just what is easiest, perhaps because the staff is already familiar with one option.

The most important message about standardization is that *more is better*. In other words, when you look around an organization, the more you can standardize—the laptops you use, the personal software you deploy, the operating systems you put in place, the programming platforms you maintain—the smoother the overall plumbing will operate.

Organizations often paint themselves into a corner by giving little thought to these internal standards, and they will sometimes end up with a hardware or software investment that is not compatible with a new initiative, which results in failure to reach the return on investment (ROI) they expected. Another way that management needs to drive accountability in the IT department is by having them present the list of standards each year and demonstrate why they make sense.

The final consideration in this step is deciding what part of the digital plumbing should be mobile. We are quickly becoming unchained from the office network, the desktop, and even the laptop computer. Just about anything we can access from a computer at the office we can now access from a handheld device. The only limiting factors are the resources necessary to write the mobile applications and the ability to reformat and secure the desktop applications into a smaller form factor. We have the ability to leverage mobile technology in a bigger way, but doing so is often not at the top of IT's "To Do" list in many cases. Because many road warriors and home-based workers carry laptops that can access company computers through a virtual private network (VPN), the need to drive applications to handheld devices has not been a burning fire. However, laptops do not fit in a purse or on a belt, so they are not handy at all times like a cell phone or PDA is. Many organizations can see big productivity gains from extending access to office data and applications to the mobile devices; determining the potential for increased mobile access at your organization is an important element in mapping your digital plumbing.

 To see an example of a digital plumbing map, visit www. velocitymanifesto.com.

STEP TWO: Upgrade the Today Maps to Show How the Digital Plumbing Should Look in Two Years

Next, a second set of maps is created that depicts how you would like the plumbing to be organized over a two-year timeline. Think of these maps as blueprints for each of the layers. Even a nonarchitect can look at a blueprint and understand what the plumbing might look like.

 Examples of two-year digital plumbing maps can be found at www.velocitymanifesto.com.

STEP THREE: Review the Maps

The technologists in the organization review the maps with the businesspeople—first the current version of the maps and then the future version of them. As a team, the technologists and businesspeople adjust and improve the maps so that everyone agrees on what the state of the plumbing is today and where the problems lie. This way, all can agree on what improvements to focus on for the future.

STEP FOUR: Develop a Written Two-Year Strategy

From the feedback and ideas that come out of this meeting, a written two-year technology strategy is developed and should include a list of specific improvements for the first year and the second year. This way, everyone is clear on what the expectations are for the coming year and has a vision of what might happen in year two. The leadership team, in coordination with the businesspeople and technologists, then determines a budget for the improvements. For more information on the process of writing this strategy, see details about the digital strategy process in chapter 6.

STEP FIVE: Review and Update the Maps and Strategies Each Year

Once a year, the plan is reviewed and updated, including the plumbing maps. This will allow you to create a rolling two-year plan that documents the improvements to be made within the next twelve months, and IT people can then be held accountable to meet these goals.

This was a high-level overview of this process, and there are many variations you can make to fit your situation. The basic concept is to use the plumbing metaphor to drive a language that will bridge tech and non-tech people so that they can agree on annual improvements to the technology infrastructure.

Understand that each of the layers of your digital plumbing must perform flawlessly in order for you to even get your email when you want it, and understand that in each layer there are a thousand things that can go wrong. Know this as well: there may be more than one hundred vendors providing pieces of the framework that supports your successful digital plumbing. One might think that the technology structure has gotten simpler as time goes on, but as fast as vendors take steps to make sure that their products fit under the heading "plug-and-play," they also add new features that affect your organization's digital plumbing and invariably cause conflicts. You would be shocked to see the full scope of the complexity of getting this plumbing in place and then maintaining it so that it does exactly what you demand when you flip up that laptop screen.

Plug-and-Play: A term used to describe hardware components that are discovered automatically, without the need for user installation or configuration.

I cannot stress enough how important it is to invest the necessary time and energy each year to reviewing your technology strategy, and making sure that you have complete agreement between the IT people and the operations side of the organization. The tendency is to do this once, then skip it for a year or two until problems begin to arise again. The truth is, the clients I have who do this consistently each year not only avoid misspending IT dollars but also trump their competitors in the market. This is simply a discipline every leader must add to his or her list of skills at this point.

CHAPTER 6

TECHNOLOGY MANAGEMENT PROCESSES

We learn the rope of life by untying its knots.

—Jean Toomer

NOW THAT YOU HAVE DRAWN UP THE MAP OF YOUR digital plumbing and identified how you want to improve it over time, we need to buckle down and move to the very practical aspects of "doing technology" on a day-to-day basis. Once you have the picture in your head of information technology as a huge set of complicated plumbing, you are then ready to move on to some practical processes for utilizing technology.

As you read the eight processes I will describe in this chapter, remember that they can be carried out regardless of how big your budget is. These processes can be implemented by small organizations and have just as much impact there as they would in a medium-sized or large organization; technology affects a five-person firm in the same way that it does in a

Fortune 500 operation. And in some cases, technology is even more critical for a small firm because it gives them the ability to compete.

These processes may completely change the flow of how your operation works. To some people, they may seem radical. To others, they will make perfect sense. Know that they have been implemented by many companies and have proven to be very effective. The goal with each process is to remove friction as you improve your digital plumbing; to hasten the return on your investment; or to help you identify technological tools that stay available, with a high amount of uptime, and can be brought back online quickly in case of a disaster.

Although the construction of technology is an artistic activity, as I will discuss later in more detail, the management of an IT structure is not. Processes need to be in place to organize and control what gets done and to tie resources to specific objectives that bring returns to the business. The processes described here do not cover the full spectrum of systems that can be put in place; they are just a few examples of how IT can be managed to get the best results.

Undertaking these processes will require discipline—they break down if only adhered to every now and then. Think of them as frameworks with which to build the structure that enables your technology artists to flourish. Without these processes, your organization will struggle to construct the plumbing that would most benefit the members using it. Leaders need to understand these processes and why they are important— just as they would understand the importance of accounting, human resources, or management skills.

You can simplify these processes for a smaller organization, and they can be beefed up for a large institution. What is most important is having some version of them in place to guide the construction of your digital plumbing.

In addition to the overview provided in the following pages, we have developed an entire suite of processes and best practices for managing technology within an organization. To see the full list of processes and how to execute them, visit www.velocitymanifesto.com.

PROCESS ONE:
THE DIGITAL STRATEGY PROCESS

Everything starts with planning. Your digital infrastructure is going to be used by every employee in every division of your organization, so employees in *all* areas of an organization must invest sufficient time in its creation. Many leaders get in trouble right off the bat when they're at the planning stage. They know deep down that technology is not helping them in the ways they need it to, yet they are so busy with daily operations that they don't invest in brainstorming how their digital plumbing should be improved. This is a serious mistake: when leaders are uninvolved in digital strategy, many different individuals or teams construct their own technology, usually in many different ways. This results in what is called a "kludge," or, in civilian terms, a cobbled-together mess of applications.

Kludge: A badly constructed collection of parts intended to quickly serve a particular function or complete a task, most frequently used to refer to crude or hastily assembled computing systems.

In order to build a mature and winning strategy, you need to develop a written two-year plan that is agreed upon by the leaders of all departments of the organization. It should list the types of applications to be used and maybe even a few features that will be brought online, but it should not go into such deep detail as to be unreadable. It is important that many people across the institution are able to read it and have a clear picture of the organization's technological goals. This strategy document will be central in guiding the improvements you will make to your digital plumbing, and it will also include many elements that would not be covered in the plumbing diagrams. For example, areas like technology training, use of social technologies, online marketing, and the connection of the plumbing to third parties are all areas that might be covered in your two-year strategy document.

In order to construct the plan, a representative group of people from across the organization must set aside a day or two—preferably off-site—to do nothing but work at a whiteboard and draw out plans that show the extensions that will be made to the current plumbing. All ideas should be measured against the potential ROI and prioritized based on the potential benefit to the company. It is even helpful if the meeting is not run by a technology person but by a businessperson who is thinking about technology as a tool that will help the company win on the streets. This could help avoid a common danger— that a technology person will limit the discussion to what he or she thinks is possible rather than addressing what is needed by the company. There can often be a huge difference between those two realities.

The plan should start with year one: list the improvements that are needed immediately. The first year is usually the easiest to strategize; make sure that the plan clearly states the exact elements to be installed in year one. Next, the year-two strategy should be constructed, and it may incorporate innovations that can't be designed at the moment. The goal is to look for high-value additions that can significantly impact the bottom line. Some people stop at two years because they simply cannot think three years ahead when it comes to technology, and that is OK—technology changes so fast that anything you write down now may be completely changed in three years. If you choose to go three years out, try to figure out how technology can be used to really separate you from your competitors. What can you do to radically change the game?

This is a rolling exercise that should be updated every year. This plan will become a task list for the IT department. In reality, there will be lots of little improvements made during the year that are not on the plan. This may happen through new versions of software that become available or through new innovations that could not have been foreseen. The plan is a high-level set of goals and should not be looked at as squashing small gains that occur along the way.

The most important thing is to agree on a strategy and then communicate it to the whole staff. Present it so that people get excited about what is coming. Let them offer their input. You will need their buy-in to implement it; software does not install itself, and data does not magically flow without someone collecting and organizing it. The best way to get the staff on board with the plan is to give them visibility of the tools *they* will have to spend hours on each day.

Once the strategy is written and agreed upon, you must have a process for executing it, which leads us to the Core Team process.

PROCESS TWO: THE CORE TEAM PROCESS

This may be the most important process of all because it provides a holistic approach to identifying technological needs and coordinating their satisfaction in healthy ways. The Core Team process helps leaders develop a team approach to making all the decisions about the technology tools the organization needs, the priorities for getting these tools built, and the allocation of IT resources needed to implement them. Unless a Core Team of people that represents the entire operation is involved in the construction and usage of technology, you will have solely IT people driving how technology is applied within the organization. This is a big mistake, and frankly is unfair to the IT department: there is no way they can know in detail how technology could help every aspect of the operation. The Core Team approach assures that technology will be applied in an orderly, efficient fashion. It also takes lots of pressure off the technologists, who will no longer have to read people's minds or guess at what they should be implementing or upgrading.

I have put this process in place with just about every client I have ever worked with, and the results have been tremendous. In all cases where the organization was struggling with getting IT done efficiently, we were able to create peace and prosperity in about three months with the Core Team process. Once you put a technology strategy in place, there is a day-to-day struggle to stay on task and set priorities on the vast amount of minutia that supports the effort. Building a coherent strategy without a process for executing it does not make much sense. This is why the Core Team process is necessary.

The Core Team is made up of a member from every major department in the organization, so it could be three people or twenty people, depending on the size of the company. The Core Team's main function is to represent the overall organization to the IT department. It meets once a month to create the dialogue that updates the rest of the company on the strategy's progress, and it also sets priorities for the technological staff so that the IT department isn't making decisions as an isolated unit. As new projects are proposed, they are documented and presented at this meeting. They then are voted on and rated as to their importance. Through these meetings, it becomes clear what IT should be working on each month and how resources should be allocated.

> **Core Team:** A team made up of members from every department in the organization that represents the whole company to the IT department.

As part of the Core Team approach, there are two important distinctions that must be made. The first is that the IT department is a service department and does not make decisions on its own as to what work gets done. Its members can suggest work to the Core Team, but they do not have the right to decide what the priorities are. This is a *very* important distinction, because left on their own, IT employees will work on whatever they think is valuable (and perhaps fun), and this might not dovetail at all with the organization's overall strategy.

The second distinction is that there are *fixes* and there are *projects*. Fixes are tasks that can be done without documentation and that should be addressed quickly as a first priority. Projects, on the other hand, take many hours or days and need to be thought through and prioritized by the team. Projects must be documented by the champion—the person who will make a case for that project to go forward—and that champion should most often be a non-IT person from the requesting group. Champions need to be willing to document a project they feel strongly about, and stay with it until it is implemented. If no one feels strongly enough about a project to be the champion, it is likely not important enough to be on IT's To Do list. This serves as a great filtering system to catch ideas that might sound good but are not truly as important as the many other projects already in the pipeline.

It is crucial for the IT department to never say no to a requested project. They should instead simply tell you how much it will cost. The philosophy here is that IT is not in the best situation to decide whether the business needs a particular project, since they normally have a view that is tainted by the specific resources they have available. It is dangerous to allow this to be the only limiting factor. The reality is an IT department can do a certain amount of things in-house free of charge (not really free) and can contract anything else out to vendors for a price. It should not be up to the IT department to decide whether using outside resources is worth the investment because that ROI decision is better made by the owner of that area.

By following a Core Team process, a number of common technology-related problems are solved:

• You avoid investing in low-value technology.

- IT departments that were previously black holes (no one knows what they are working on, when they will be done, or how their resources are allocated) become more transparent.

- The broader community is united with the IT department, which helps ensure that the digital plumbing is built correctly.

Experiment with the Core Team process for ninety days and you will experience its power firsthand.

To download a document with more information on how to implement the Core Team process, visit www.velocitymanifesto.com.

PROCESS THREE: THE IT RESOURCE ALLOCATION PROCESS

The process for resource allocation should follow a simple model. Projects are documented on a Scope of Work (SOW) document. The SOWs are approved (or rejected) by the Core Team. As the projects are approved, the IT department works with the champion of the project to define it in detail and make an estimate of the number of hours needed to complete it. Then the project gets slotted into the calendar based on the status of other projects. The IT people do not make this decision; the Core Team does. If the Core Team decides that nothing can be delayed to make room for the new project, a member of the team and the project champion go before the decision makers to ask for the extra resources. They will make the case for the project using the SOW, which has a section that defines the expected ROI of the project. They will either get the resources or they will not.

> **Scope of Work (SOW) Document:** The document the Core Team uses to set timeline, deliverables, and expected ROI for a proposed project.

This is a simple way to ensure that the IT department has sufficient resources without the decisions being political or coming about as a result of historical inertia. This process also keeps the IT department focused on execution and not the management of resources.

PROCESS FOUR: BUILD VERSUS BUY

The process of deciding whether to build or buy technology applies only to software, since most people don't care to build their own hardware these days. Although small operations may not be able to afford custom-written code, larger organizations often do have the resources to build their own software tools. If you do have these resources, you must decide between buying off-the-shelf software or building your own from scratch. This process must be well defined because down each branch there are important steps. So let's start with the big question and then deal with each option separately.

When faced with build-versus-buy decisions, many companies make the decision overly complicated. Software is just a tool, and it's pretty easy to figure out what you need from that tool and what it will cost to have it implemented. The problem in real life is that many IT people have a vested interest in the outcome of the decision, and because executives often do not have the knowledge to challenge the advice of their technical people, they just go along with what the IT people suggest. Here is a set of basic questions that will help you come to a knowledgeable decision:

1. **Have we researched every product on the market that might do what we need?** In reality, people sometimes do very little research before investing huge amounts of money.

2. **Are any of the available products set up so that we can buy the code and modify the application at will?** Limiting yourself by purchasing software that you cannot customize is flat-out dangerous.

3. **Can we build something in-house that will have the richness of features the purchased software has, and will we need that richness at some point?** Building rich software is going to cost not only money but also time. If there is something you can buy that has tons of features—perhaps even more than you need—you must consider how long it will take you to construct something at a similar level of sophistication.

4. **Is this application going to give us a serious competitive advantage if we build it ourselves?** There are times when building custom software that no one

else has will give you a big-time competitive advantage in the market. If the reward for building something customized is worth the risk, it makes sense to build something that's a generation ahead of your competitors.

5. **Have we proven we can build custom software in-house?** None of the previous questions matter if you have never been able to successfully build software in-house. Don't bet on a losing hand if you have little confidence that you can build software from the ground up. Get a vendor to build it, or buy something off the shelf.

6. **What are the respective costs of each option? (Whatever you think they are, double them to get to what it will probably cost in reality.)** There are times when building custom technology costs less, and times when it costs more. The price difference must be weighed against risk and the distraction of building in-house.

7. **How much time will it take to get each option live? (One may take much longer than the other.)** Time is a big factor. The longer you go without the software capabilities, the longer you live with the inefficiencies you now have. Be careful of internal developers who give unrealistic delivery dates just so they can do a pet project in-house.

8. **Can we build something in-house that can be managed and upgraded by new staff in the future if we lose the developers who will code the**

original application? In other words, make sure you factor in who is going to do the future support. I have seen custom code built in-house that cannot be used once the builders move on. This leaves you buying software years later because you should have done it in the first place.

9. **Does writing a custom application add value to my company because we will own the intellectual property?** In the world today, companies are beginning to be able to list custom software as an asset unto itself. If you can write something and then own the intellectual property in a way that creates clear value, it makes sense to build rather than buy.

10. **Is this software at the core of what makes us succeed or is it just a back-office tool?** The more central to your organization a piece of software is, the more you should be willing to custom build it. Functionality that is not at the heart of what you do should more likely be bought.

By the time you are done answering these questions, you will be able to determine which option gives the organization a faster ROI, and which has less risk of failure.

PROCESS FIVE: THE SOFTWARE IMPLEMENTATION PROCESS

Implementing software is one of the biggest sources of frustration for many executives. There is always a high level of excitement on the front end about the advantages of a new piece of software. Then reality sets in as people who already have many responsibilities have the added burden of getting a new software system up and running. The result of implementing new software without a structured plan and loads of discipline is normally blown deadlines and blown budgets. The following is a basic process that can be used to keep things on track during implementation. It may need to be modified depending on the size and complexity of the project and is simply meant to give a sample of the kind of process that needs to be in place in order to have a shot at success.

The first step in a productive implementation is to get the project off on the right foot by setting up the vendor or the internal staff to succeed. (In this process, we will consider anything that is written internally as being installed by a "vendor," since the process does not change when you are using internal developers.) There might be two vendors involved—one that produces the software and one that resells and implements the software. The installing vendor must perform as they have promised in the selection process or the implementation may go seriously off track. The cost to the company will be the wasted time and the extra money it might take to get the implementation back on track. The following is a list of specific steps that must be completed in order to ensure that the vendor performs.

STEP ONE: Contracting with the Vendor

If you are using outside vendors, signing a contract is a must. When putting the agreement together with the vendor, it is a good idea to try and break up the "sale" into pieces. This is only necessary if you have no prior experience working with them. It is best to start by having the vendor do a small piece of work to see how they perform, and maybe even implementing the software through a pilot phase. This not only limits your exposure in the agreement but also keeps pressure on the vendor to perform, because if they blow the front half of a job, they will not get the second half. The bottom line is that it is not wise to sign deals that assure a vendor the sale of a large project; then the vendor owns the whole process and has all the leverage. What follows is a list of specific terms that are good to get into any contract in order to incentivize good behavior:

- Penalties for missed deadlines

- Fixed-price agreements for implementation

- Service level agreements (SLAs) if the software is hosted

- Language that assigns a specific team, which cannot be changed without client approval

- A schedule for meetings and reporting: if the project is medium-sized to large, there should be a weekly meeting at a set time. Meeting attendees will review what has been accomplished and then write a list of any further problems to be solved or issues to be decided on.

STEP TWO: Problem Resolution

During every implementation, there will eventually be issues. These can range from projects that get behind schedule or go over budget to software that simply does not meet the needs it was purchased (or created) for. Problems must be dealt with quickly and escalated up the management chain to the vendor as soon as possible if solutions are not found. The biggest mistake you can make is to ignore problems or wait too long to deal with them. It is critical for the vendor to know you want any issues solved in a timely manner.

STEP THREE: The Implementation Team

For any project, regardless of its size, there needs to be a team that is matched to the job. In other words, the team could be two people or ten people depending on the size of the effort and scope of the work involved. The purpose of this team is to be responsible for following the steps listed here and to "own" the implementation. It is best if this team has members from each of the departments or groups that will be impacted by the software.

The team should be formed immediately after a decision is made to buy or build a new piece of software, at about the same time the vendor is selected. It will interface with the vendor, drive the implementation, and provide any needed input. This team will also be responsible for helping with the "go-live" and reporting on the success of the implementation afterward. If problems arise on the project, this team must get them resolved or escalate them with the proper parties.

STEP FOUR: Communication with the Company

During the implementation, communication with the proper people in the company at the right time is extremely important. Management needs to be updated on progress regularly and at short intervals, and they also need to be notified of any serious changes to schedule or budget. Additionally, the *whole organization* needs to be updated from time to time as to the progress of the install, especially if the software is going to have a wide impact on the employees.

STEP FIVE: Regular Updates

The team will need to give project updates to the organization, and it is best if these updates are done on a predictable and consistent basis. Depending on the length of the implementation, this could be weekly, biweekly, or monthly.

STEP SIX: Timing Notifications

Make sure you provide information to key members of the organization in advance of such important events as a go-live or a soft rollout (defined on the next page). Also make sure you notify the necessary people when there are issues. Do not let the water cooler be the place where information is passed. This is true for good *and* bad news.

STEP SEVEN: Testing

This step is very often given very little attention or ignored completely. Testing needs to be built into an implementation at all points of the process. The more testing you do, the more likely the go-live will be smooth and you won't embarrass yourself with lots of post-installation problems. Here are the three major types of testing that should be done during a software implementation process:

Unit testing. This is testing of specific pieces of the software; it can be done as parts of the system are put in place. Unit testing should be done first by the programmers, then by the implementation team, and then by users. Because it assesses functionality only, unit testing can be done with a small set of test data.

Application testing. This is testing of the entire application for viability, stability, and throughput (i.e., the rate at which data is processed). This level of testing should be done with a full set of data. The programmers, the implementation team, and then the super-users should do testing.

Final testing. This level of testing should be done by users who are completely unfamiliar with the software. Its goal is to see how standard users view the system, and what they can find to break.

STEP EIGHT: Rollout Options

A failed rollout will taint the entire project for months. For this reason, it is extremely important to be intelligent about how your team rolls out a new application. The implementation team must spend sufficient time thinking through the best way to roll out a system so it has the best chance of success. Here are a few rollout options:

Full rollout. This is a rollout to the entire company, and each employee will run the application at the same time.

Soft rollout. In this rollout, the implementation team picks out one group (typically the group most prone to making mistakes with the software) to test the system on; it is rolled out in stages from there.

Excluded rollout. This is a rollout to the majority of the organization *except* the group that would be the most problematic. Saving them for last lets the team customize the rollout to the group that it knows will be difficult.

STEP NINE: The Go-Live

The actual go-live is the stage in which you run the system as a regular tool for the organization. This is also the stage where problems can become a huge distraction and have very real costs. The go-live must therefore be well thought out and executed flawlessly. Make sure you have the following areas covered to avoid unnecessary difficulties during the go-live:

Communication. Always *over*-communicate with the staff on what you will be doing, when it will happen, and what is expected of them. This communication must start a good while before the go-live so that people have a say in the timing and are prepared for the time they will need to set aside for training.

Staffing/support. You simply cannot be understaffed during the go-live. This means making sure the vendor is on-site, or at least has plenty of people available to help with questions and problems. You also need to have people from your implementation team ready to help new users until they are comfortable with the software. This normally takes only a few days, but you must prepare for it.

Training. Please do not skip or go light on this step. Users must be trained in advance of a launch. You cannot install new software, pick a date to go live, and then just hope people will pick it up without any trouble. You must train them until *they* feel comfortable they are ready to go.

Problem resolution. You must be prepared to solve technical problems after the go-live. There are always issues that will come up in the live environment, and you need to be prepared to handle them quickly and cleanly. It is naive to assume that the first few weeks will go smoothly. Software is just too complicated to foresee everything that might go wrong.

STEP TEN: The Return on Investment Study

Once a software implementation is done, the last duty of the team is to write up a report that measures how successfully the system is meeting the original goals. This report is given to management so they can learn whether good decisions were made on buying the software, and whether the software is indeed providing the ROI that the team believed it would get.

An important note: I am floored by how many organizations will spend what for them is a huge amount of money on technology and then not do the math to see whether the investment is paying off. Don't skip this step; it will help you know whether you are making good or bad decisions.

PROCESS SIX:
THE SINGLE-POINT-OF-FAILURE ANALYSIS

This process is fairly simple to execute and helps the organization understand the places where the plumbing is most likely to break and what the impact will be if it does. It helps close the gap between what IT and management see as the right way to construct the digital plumbing. In many cases, the executives have no idea where the danger points are and then are surprised when the day comes that the organization goes "off the air," so to speak. They blame the technology people, but the technology people are livid because they felt they had mentioned the danger and no one listened. The bottom line is that both parties need to make a conscious decision as to the level of risk the organization should take when it comes to the consequences of various failures.

The first step is to create a spreadsheet that lists all possible points of failure down the left-hand side. This includes issues like single routers into the office that if broken would take the Internet down for everyone, and single servers that when down render applications unusable. A single Internet service provider (ISP) for your building means that everyone will be down when that ISP is down, and they *will* go down someday. Thought must also be given to staff-related points of failure: Where do you have a single person who provides support or holds knowledge that would be lost were they to be off-site or were something to happen to them? It often takes a few people brainstorming together to really complete the list of potential single points of failure.

Single Point of Failure: Any part of a system that, if it fails, would bring down the rest of that system.

The second step is to add a column that describes the impact of each failure. In other words, what will your organization not be able to do if this condition arises?

The third step is to add a column that describes how long it would take to repair the damage that could result from the failure. In some cases, this will be unknown. For example, if you have a single ISP, you are at their mercy as to when they will restore service. But if you lose a router, you can guess at the likely turnaround time to replace it.

The fourth step is to add a column that establishes a low-expense fix for the situation. This will not be the most professional solution, but it will be one that improves the circumstance. For example, if you do not have a backup power supply from a diesel generator to keep your servers functioning in a blackout, you could buy a portable generator and run extension cords out of your data center to keep things up and running. This isn't the preferred method, but it will work in a pinch and, of course, costs much less than the truly professional fix.

The fifth step is to create a column that describes the full professional fix and the price associated with it.

The last step is to have a meeting with the leadership to review this single-point-of-failure analysis and make conscious decisions as to what risks you are willing to take and what investments you might want to make to get rid of weaknesses in your digital plumbing.

PROCESS SEVEN:
THE DISASTER RECOVERY PROCESS

Disaster recovery processes have become a frequent topic of discussion over the last few years. Some organizations go to the extreme in making sure that under *no* circumstance will the business be impacted in the event of a disaster. This is easier to do if you are part of a large institution with lots of resources and offices around the country. Many smaller organizations have not dealt with this subject because they just have too much on their minds as they try to survive day to day. Many people seem to think that investing time in a disaster recovery process is not valuable because the odds are so low that anything will happen. They're OK with just accepting the risks. Before we go any further, maybe we should step back and examine why this is becoming a hotter topic these days.

A few factors have raised the importance of having a business continuity plan when a problem arises. The first is that organizations have become more dependent on technology to operate, so if the power goes out or the data center is physically damaged, it often means the business comes to a halt until the equipment can be brought back on line. The second factor is that customers have more ability to move their business to a competitor; in this increasingly competitive market, loyalty to any one vendor has been decreasing for years. So if a supplier of goods or services is off the air for more than a few hours, it will not only lose some business—it may lose customers for good. The third factor is that the expectation of quick recovery from technical difficulties has increased. As larger organizations invest in disaster recovery plans and become more sophisticated in how they maximize the business's uptime, customers have become less patient with smaller vendors who have issues that cause downtime. It is not unusual now for a large customer to threaten a smaller vendor that has just experienced a disaster by telling the vendor they will move their business if a disaster recovery plan is not put in place. The large customer reasons they cannot depend on the smaller vendor if it is not willing to minimize the risk of downtime, especially when others in the market have implemented advanced disaster recovery processes.

Here are a few suggestions as you consider the best process to help your company recover from unforeseen events:

1. **Assemble a team** from across the organization that will build the disaster recovery plan. You should be on the team to take action should any difficulties arise. Do not restrict membership of this team to just a couple of people, or just the IT department. The whole business will be impacted in a disaster, so the whole business must be represented when building a recovery process.

2. **Review each division** of the operation, especially the IT department, and discuss how that division could be built back up in the case of total loss of either the home office location or any of the other locations around the world.

3. **Define responses** for issues of varying severity:

- A half-day power outage

- A bad storm that cuts off access to the building for a day

- A small fire that damages part of the building

- A tornado that destroys the whole operation

4. **Create To Do lists** that will be used in the event of each situation so that tasks can be handed out and executed quickly.

5. **Have recovery meetings** every six months to review any changes that need to be made. Practice by naming a scenario and having everyone on the team list his or her first five moves.

PROCESS EIGHT: THE TECHNOLOGY RESEARCH AND DEVELOPMENT PROCESS

I have yet to see an organization that has developed a structured approach to studying new information technologies and functionalities as they come on the market. I guess this is so often overlooked because technology innovation is coming so fast that most leaders don't believe they can even stay current. In addition, in the minds of many leaders, finding new technologies equates to spending more money and using more resources—or they regard new technologies as toys or shiny objects that will just distract people from the tasks at hand. And while I understand all of this reasoning, I still believe that a powerful leader needs to be aware of all the tools that can help an organization prosper.

There are a couple of ways this process can be built into the culture of an organization. The first is to appoint someone in the technology department as the person responsible for studying and testing new things. When she finds something that looks promising, she takes it to the staff member that would be most interested in it. This way there is a continual flow of options for possible technological improvements being shared within the organization.

The method that I prefer is to pick one person from each operating group and have each of them dedicate 10 to 20 percent of their time to researching new technologies or functionalities. This seems to work better in most cases because the people doing the research have more specific knowledge of what would help their department than an IT person would have. If this task appears on the employee's annual review—or if the employee is rewarded for finding technologies that turn out to make a significant ROI—it is amazing how many ideas they bring into the organization.

Both of these processes ensure that the organization does not have a set of stale tools at its disposal. I have long observed that companies sometimes exist in an ebb and flow with technology as they feel the pain of a needed upgrade, implement new tools, then succumb to software fatigue and stop looking at any new technologies for a couple of years. In this day and age, checking out of the game for two years could cause irreparable harm, so I do not advise accepting this ebb-and-flow approach. It is better to know what is available to you and consciously choose to delay experimentation than to remain ignorant of your options.

Don't think this process is the least important because it's the last one in the chapter. If technology is a set of weapons that can be used to win in the battle of the marketplace, then it makes sense to understand what your inventory should look like. No one enjoys learning about a great new technology from a competitor.

We can reduce all the processes listed in this chapter to one simple progression of thoughts:

- You need to have a way to decide what your organization's digital plumbing should look like.
- Next, you have to know what to build, and in what order.
- Then there has to be a process in place to actually implement the projects.
- As the plumbing is constructed, you have to constantly review the stability of the framework and harden the dangerous spots.
- Finally, you will want to figure out how to make the plumbing as easy as possible to repair if a section gets destroyed.

None of these processes will be implemented without leaders who understand the value of organized systems and strong digital plumbing. I promise you, I am the last guy to want more bureaucracy. I hate details and paperwork. But I can also promise you that if you do not put processes like these in place and fight to keep them on track, your business results will suffer. There is too much at stake for any organization to exist in technology chaos.

CHAPTER 7

THE DIKW CHAIN

*The greatest obstacle to discovery is not
ignorance—it is the illusion of knowledge.*

—Daniel J. Boorstin

I LOVE THE ABOVE QUOTE FROM DANIEL J. BOORSTIN
because it so eloquently portrays what I sometimes hear from
businesspeople I work with. They believe they see all the
important data and have every important analytic their business could possibly supply. In other words, they think they
have 100 percent of the knowledge available. I call this delusion "data hubris." These leaders really believe they capture all
the necessary data, that their reporting is as good as it gets, and
that they can identify all the important trends in their business
using the means they have today. In other words, they have no
idea what they do not know.

Data Hubris: The delusion of believing that you currently capture all the data necessary to effectively lead your organization, without analyzing what key knowledge you could be missing.

Once you have your digital plumbing in place, you have to focus on the data that is flowing through it. It isn't enough to just get data efficiently. You must gather a large amount of data from three different sources (internal, constituents, and third parties) and know what to do with it.

If you want to be a high-velocity leader, you must understand what it takes to have the *very best* visibility into what your organization is doing and what the performance trends are—in real time. If you have data hubris, you will not even be motivated to find out where you are missing important facts. You simply cannot confidently keep pace unless you can clearly see the current status of the machine. The more robust the information flow you develop, the more you can move to predictive analysis, since you will have a base of knowledge that allows you to extrapolate what the future might be. Speed requires more information and the ability to adjust to conditions that come up unexpectedly—just compare the dashboard of a car to the dashboard of a jet airplane. Rapid but wise decisions are predicated on a solid flow of high-quality information.

Once the digital plumbing is flowing, we must improve what flows through it. Raw data is useless unless it gets turned into information. Information is useless unless it gets turned into knowledge that gets acted upon. Knowledge can be dangerous unless it is tempered with institutional wisdom. Every organization is immersed in raw data, but most do not transform that data to a state in which it can be leveraged. Data is generated and then abandoned to live in databases for decades without ever being conscripted to help executives make wise decisions. With the right perspective, executives can ask the right questions and hold IT departments accountable for turning the data into useful dashboards and analytics that will make a huge difference in the directional choices leaders make. Since knowledge is power, more knowledge is a creator of velocity, and we are still in the dark ages of acquiring and utilizing the full breadth of data available.

There is a natural upward flow in value from **Data**, to **Information**, to **Knowledge**, and finally to **Wisdom**—these are the concepts that make up the DIKW chain. This series of words describes a progression that happens thousands of times a day, and we normally call it "having experience." By understanding how to drive our intellectual assets up this value chain, we can make better decisions about how we apply resources. Better application of resources translates into winning in the market. There is tremendous value in applying this concept to your business, and all it takes is looking at what you are doing with a fresh mind-set. Let's examine each of the stages individually, and then put them together as a process.

DATA

Data is the plural of the word *datum* and refers to individual facts or pieces of information.

In the last four decades, we have learned how to capture data in huge quantities and store it safely. We have built storage

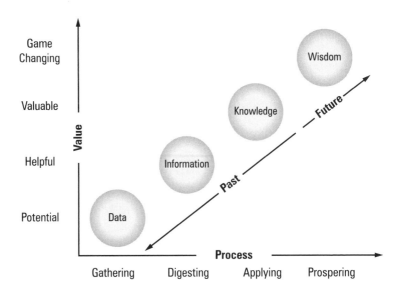

applications called databases and many data-input devices like scan guns, UPC readers, and keyboards. We have even been pushing the data input out to customers and clients, who are now filling out forms on the web to save us the trouble. These databases have become very powerful platforms for storing raw data in *huge* quantities. They are easy to modify and edit and have solid reporting tools for extracting information. Databases will continue to expand in functionality, and they already provide more capabilities than most people understand. The term *data mining* has become well known because people now understand there is gold in our databases that we can find if we just go digging for it.

A database exists beneath just about every software application we use. It is the archive that stores every piece of information needed for an application to run. Through these applications, we fill our databases with *huge* amounts of raw data. In the past, this data existed on paper and was stored in file cabinets. The ability to store all this data electronically provided huge savings in storage space, file cabinets, storage boxes, and retrieval costs. Unfortunately, some people stopped at the savings and did not take the next step of actually leveraging this electronic data store. Others had a clear picture of the value that sits in a database and invested wisely in business intelligence tools that can sort and present this data in many ways. They have even taken the next step and built "rules engines" that actively monitor the data store for specific conditions and then trigger actions that help solve problems in the organization without the need for human supervision.

> **Rules Engine:** A software system that enforces a set of rules by monitoring an organization for certain conditions and then taking action when any of those conditions comes about, without human oversight.

There are two very important points to understand about the state of data management today. The first is that we need to do a much better job of mining the data we already have. The second is that we need to gather more of it. For example, most companies do not go out of their way to learn everything they can about their customers. They record only the bare minimum necessary to complete a transaction. They assume that asking for any information more than that will seem intrusive. But the fact is most people are glad to share lots of information about themselves as a trade-off for more targeted marketing they would find valuable. There is a segment of society that feels it is invasive for companies to ask for more data, so you must be sensitive to this and never force people to provide data they don't want to give. The other issue is the time it takes to provide data; when your customers are in a hurry, get the bare minimum. When they are not, you have the chance to collect more.

Let's use a typical bank as an example. When you open a checking account, they will have you fill out some forms by hand and then they will open an account with the six or so pieces of information you gave them. They never ask other basic questions, such as: Do you have kids? Do you have a relationship with another bank? What loans do you have with other institutions? How long have you been in town? Where do you work? What is your birthday? What is your email address? The list goes on. Now, most people do not love to be asked one hundred questions in a row, but answering a handful so their bank knows them better as a customer and can work with them more effectively is usually just fine. Let's say the bank actually records your answers in a database and then uses a simple piece of customized code to send you emails that offer services or contain a birthday coupon. You can see pretty quickly how the bank could actually live up to its promise of being a friendly hometown institution (even if it's a billion-dollar national conglomerate).

For an example of the lack of streamlined data collection processes, let's look at a hospital. It is very common for hospitals to advertise how much they care about their patients—yet notice the check-in process for most hospitals and watch as they have you fill out form after form by hand and, in many cases, ask for the same pieces of information over and over.

What bothers me more is what they do *not* ask. As a human being who is about to check into a place where I will be trusting other people with my very life, I think it makes sense for the hospital to actually ask a few questions that do not concern how they will get paid or who my employer is. In other words, in order to fulfill their mission of being my attentive, local healthcare provider, they need more data on me. This data could be stored in the same database they already have by simply adding a handful of fields. Then that data can be passed to nurses and doctors alike so I can be given more customized care; that is, if they really do intend to provide that kind of care. For example, why not ask about my level of fear at being in the hospital? How about my expectation of how long I will be there? It would be nice for them to know whether I prefer peace and quiet or lots of activity rather than choosing for me. They could also ask how often I would like to be updated on my condition. This list, too, could go on and on.

Some of these questions may seem strange, but when you are a patient, these are the exact areas that cause you the most frustration with the hospital staff. The reason most hospitals do not understand the need to gather this kind of data is because they don't want to respond to it. If they chose to, they could use technology to ease the burden of meeting patients' expectations by automating some of their activities through simple notifications triggered by patient data gathered at check-in.

Lest you think I am being too hard on hospitals, this same example could apply to just about any business—including yours. We often do not gather all the useful data that would allow us to create a better customer/client relationship. When we don't do this, we fail to wisely use the data we have. Typically, the larger the organization, the more data that gets generated; unfortunately, the harder it also becomes to use it valuably. The smaller the organization, the less data there is.

Small organizations also typically have fewer resources to mine what they do have.

Let's move on to what happens when we do put data to good use.

INFORMATION

Once data has been pulled up and put into some type of context, it grows up to be information. For our purposes, we will differentiate information from data by saying that information is *data that has been put in a specific context for a specific purpose.* For example, millions of times a day, Walmart runs products through a bar code reader as they check people out at the registers. A database stores information on every single item that was sold, where it was sold, and what was charged for it, along with a few other pieces of information. In its raw form, that data is not meaningful—it is a monstrous amount of numbers and text. But once a manager asks for a report that puts this data in context, it becomes useful information. For example, the manager might ask for a report that shows the percentage of grocery items sold versus clothing items sold. In addition, the manager might ask to see a chart of this relationship over the last year. That chart might show a trend toward groceries becoming a higher percentage of the sales as compared to clothing. This could be a problem if groceries are less profitable than clothes are.

In reality, there is a treasure trove of information embedded in the sales data of any large store, and it is just waiting to be quarried. Compare the individual store's information with numbers from other stores across a chain and you have even more information. Add to that the ability to check one store's numbers against industry numbers and even more useful information will be revealed. All of this information is available because we have stored a huge amount of data in a database and have the ability to report on it instantly. Such is the case in almost every organization in business today that uses a computer for sales transactions.

There is a gap right now between the value of basic data and the lack of value in the information pulled from that data. This gap has grown because most managers and executives are not proficient in the use of reporting tools and business intelligence software. They simply do not know the capabilities of the tools available today. In addition, many IT people will not go out of their way to teach non-IT employees about these tools. This often occurs because IT departments are overwhelmed, although in some cases they may purposely withhold this information from others to increase their own perceived value to the operation. If you are in a management position of any kind, this is your wake-up call that you have likely been flying without nearly the level of visibility you should have. Raw data is not that useful to you. Turning that raw data into useful information will help you make better decisions, which will result in you reaching your goals. If you do not have every single analytic you need to do your job, it might be good to take a closer look at what is going on with your technology staff.

The most difficult part of this stage of data evolution is simply getting people to imagine what they need, since many of them may believe they have everything that can be gotten. It takes some creativity to imagine what analytics could be valuable for a specific role. Many people simply do not have this level of creativity without being led to it through a series of the right questions.

KNOWLEDGE

Once information has been created from raw data, it can be translated into institutional knowledge. For our purposes, we can define *institutional knowledge* as information that is embedded in the brains of managers and leaders, gathered over time, that is ready for application. Information on its own is simply organized data. It may say something interesting, but that does not mean it is understood or applied. But once it becomes institutional knowledge, it can be applied in order to drive improved results. It can also be multiplied and interpreted by a large group of people, so that many employees can alter decisions based on this new information. This is where our digital plumbing becomes very important: once we have created information from data, we need the ability to quickly turn it into institutional knowledge to be used for the greater good.

Institutional Knowledge: Knowledge collected by a set of individuals in an organization, which must be effectively transmitted and understood to be valuable.

What a waste it is for one person to have mined interesting information that is then held solely in his or her own mind. And this type of waste happens every day, in every organization. In some cases, people simply do not want to share valuable information because they know keeping it to themselves can give them an advantage inside the organization. In other cases, there are no tools within the technological infrastructure for sharing it effectively. People are reduced to emailing information to each other, and although this can work, it is not the preferred method. In the best-case scenario, a person would discover interesting information and have an automated method for creating a list of people who might be impacted before pushing the information to them automatically from that point on. Processes like this are what create valuable institutional knowledge.

In a world where businesses are getting more competitive every minute, it becomes harder to have an edge—but it also becomes much more important. An easy area for a company to get ahead is in its ability to move from data to information, and then to infuse the information into the institutional knowledge base. Then it can be leveraged in the marketplace against the competition. Otherwise, you'll end up watching as your competitors make decision after decision that may make no sense to you at the time, but result in one of them someday buying your assets in an auction.

WISDOM

The final step in this process is the creation of institutional wisdom. Knowledge is a powerful thing, but it needs to be tempered with wisdom—even solid knowledge can be applied in ways that can be disastrous. Institutional wisdom is the ability to apply knowledge with the discernment that can only come from human instinct and observation. This is not an area computers have yet advanced into. We can build software-based knowledge systems that can simulate a *percentage* of the decision-making capability of an expert, but not the full capability. There are simply too many variables in most decisions for us to fully depend on a computer. This is not meant to denigrate the power of software-based knowledge; there are many circumstances in which results can be improved by building software-based models for sorting through a large set of variables and predicting a correct response. However, we must acknowledge that in most cases, we still need human input at the end to observe the overall picture and approve or expand on what a computer might suggest.

Let's look at the example of software that predicts possible causes of health issues based on a collection of symptoms. We can gather lots of data on a sick person. We can run tests, collate the findings, and sort out the variables that are outside of the norm. This helps us create a set of information that might be useful. Let's say we have a person with a high fever, jaundice, and tenderness in the lymph nodes. We put these symptoms in a knowledge-based software system and it comes back with the speculation that the person may have hepatitis. At this point, we might run specific hepatitis tests and then prescribe medicine to counteract the disease. However, if we instead apply some wisdom before running these advanced tests, we may discover that the cause is actually liver failure caused by chronic alcohol abuse. This we can learn by simply noting the sweet smell of alcohol rolling off the patient.

The same dynamics exist in the corporate world. We should work hard to turn data into useful information, and then embed that information as institutional knowledge across the organization. Before we act on this knowledge, we should always integrate it with institutional wisdom so that we do not misuse the information. You can see wisdom in action if you watch successful businesspeople in a down economy. Although their profits might be falling, and although the market looks negative for the operation, some aggressive operators will use the situation to expand market share and invest heavily in consolidating new territory—even at the expense of losing money during the recession. All of the information and knowledge they have may be saying that the business is going backward, but wisdom says there is opportunity to make great gains while others are reeling.

You can find much more information regarding the DIKW chain on the Internet, but before we move on, there are a couple of things to note concerning the transformation of data into valuable wisdom. The first is the concept of creating a rules engine that sits on top of a database, constantly watches for preprogrammed conditions, and then spawns actions. This can be one of the most powerful and cost-effective things an organization can do to drive improvements. Since the underlying database holds a list of all transactions and many of the communications that are going in and out of an entity, it is possible to build a set of rules that are triggered when a preset condition is met. A simple example is a manufacturer who wants an automatic email sent to management if any of the following occur: a sale that is lower than 40 percent gross margin, a return of goods that is over $5,000, or week-over-week sales growth that rises more than 25 percent. These are only a few examples of rules that should be part of a set numbering in the hundreds. In so many cases, we choose to fly blind simply because we do not understand how easy it is to build sets of rules that constantly monitor new data.

By surveying every manager in an organization and asking them all to develop a list of rules and controls they would love to have in order to ensure that their department is running as it should, you can create hundreds of rules that make tons of sense. You can simply ask them what condition they would like to know about, and then ask how they would like to be notified—email, spreadsheet, printed report, daily, weekly, monthly, etc. Forget about what you think your software can do at the moment. Just generate a list of all the rules and controls you would like to have. Then ask your IT people to implement the list; you will be surprised by how easily they can get these rules in place.

Patterns swirl around inside of organizations constantly. Many of them may not be recognized at all, and others get recognized too late. At the moment when managers can begin to wrap their heads around how to ask for tools to monitor these patterns, a whole new wealth of knowledge can be accessed. The difficulty in getting to this knowledge is that it often takes a person who is not only steeped in the daily operations of the organization but also technical enough to understand how software might be written to identify or highlight important trends. The right climate for the creation of this type of person has arrived: the generation of workers coming up today has been exposed to enough technology that as they learn the crucial patterns of an organization, it will be more natural for them to request monitoring and know exactly how to structure their requests to an IT person.

Some organizations are already beginning to leverage these more sophisticated methods for mining underlying data. These pioneers are being rewarded with information they, in turn, are leveraging to separate themselves from competitors. They are finding that the incremental investment to add these capabilities to their existing IT structure is minimal, and that the rewards are substantial.

Deep inside the data is the truth about your business. The data does not lie. It documents and archives our transactions, preferences, results, and activities. Sometimes we do not want to know the truth; we want to delude ourselves into believing our own version of reality. This willful ignorance leads to pain. There are two ways we delude ourselves: actively and passively. In one case we purposely do not mine the data because we don't want to face the truth. In the other, we are truly ignorant that the truth can be brought to light. Either situation is tragic.

CHAPTER 8

BUSINESS INTELLIGENCE

We don't receive wisdom; we must discover it for ourselves after a journey that no one can take us on or spare us.

—MARCEL PROUST

THE CONCEPT OF BUSINESS INTELLIGENCE (OFTEN abbreviated simply as BI) has gotten to be a hot topic lately—it is the term of art for the process of turning the DIKW chain into something practical. The business intelligence field is still developing, so I will aggregate some of the existing ideas on it, add a few of my own, and simplify the language so you will be left with specific processes that can help guide your organization as it matures data into useful knowledge and wisdom. The value of learning to build better visibility into your information and creating useful analytics cannot be overstated. Most leaders mistakenly believe they have great access to information and trend data, but in most cases they simply don't understand what they aren't getting—because they don't know what they *could* have access to!

Better data analysis allows you to increase the velocity of your organization. There is a good reason we use the metaphor of a dashboard when talking about aggregating analytics and data onto a computer screen for a leader: as the velocity of the economy speeds up, and as organizations get more complicated, spread out (virtual), and have more moving parts, a leader needs a dashboard to understand the speed and direction of the organization. In order to lead effectively, you must be comfortable with the pace of the organization, and to be comfortable, you must know that the underlying performance and dynamics are within acceptable boundaries.

Business Intelligence: The process of gathering deep analytics, trends, and intriguing bits of knowledge from a large body of information, and then creating methods for seeing it in ways that give it valuable context.

Business intelligence can provide you with much better data and analysis as you move the organization forward and determine its overall health. Many organizations stumble ahead, partially blind, doing things as they have always done them while still breaking even or making a little money. They subscribe to the "I am still alive so I must be OK" theory. But if they really understood the underlying numbers, they would see that they are starving—they just don't know it yet.

At the very foundation of business intelligence is the understanding that there are three piles of data you can access to provide better analytics. The first is the data you already have in your plumbing. Truth be told, I would guess that most organizations utilize only 60 percent of the data they already have. For example, the stored data from transactions they do with customers, or the web trends data on who has come to their website and what they have done there. A lot can be done just by working harder to create value from the data you are storing now.

The second pile is all the data you could have if you asked for it. I have found that I can put a client in a room and ask them to think about all the data they would love to have in order to sell more, to create tighter relationships with customers, or to have better visibility into trends, and without struggle, they can identify fields and fields of data they could actually get if they would just ask for it (or purchase it). So many leaders fall into the trap of thinking they can only have the data they have right now, instead of opening their minds to what is possible.

Think about the small amount of information a real estate agent gathers when she is trying to find the right house for a client. She usually asks these basic questions: What price range are you looking for? What school district do you want to be in? How many bedrooms do you need? But without much effort, she could also ask a few more questions that would provide a better set of data by which to help find the ideal house: Do you have pets? How deeply do you sleep? Are you sensitive to temperature? Do you like to socialize with neighbors? Do you have any allergies? I could go on and on. Each of these questions has a direct bearing on what kind of house and what kind of location would be a good fit. But for the most part realtors do not ask these questions. Why not? Because they never have before, and inertia is a powerful force.

The third pile of data is third-party data. This is all the data that is now being provided in streams, in many cases at no cost, over the Internet. With all the programming tools we have today, we can mash up data streams—or overlay them—to see combinations of our own internal data and third-party data. A simple example of this would be combining data from Google Earth, census data, and your organization's sales data. By doing this, you can create maps that show you where your sales are coming from geographically and the demographics of the people buying them. Using this pile of data takes creativity, and that is precisely why many organizations are not tapping it. The IT people know how to do it, but often they do not share examples or even know exactly what the leaders would want to see. The leaders might have an inkling of what they would like to see, but they have no ability to express it to the geeks.

Once you have the mental picture of the three piles of data waiting to be tapped into, you can move on to the practical methods for developing a business intelligence strategy.

CUSTOM REPORTING IMPROVEMENTS

The simplest method for developing mature, useful data is to use either the reporting systems already built into your databases or other widely available third-party reporting tools.

Really good programmers will tell you that if the data exists in the correct format, a report can be created that will supply almost any view you want. If you can think of it, they can do it. Good programmers not only have the ability to simply pull the data and express it in the format you want; they also have the ability to write code that will filter the data into discrete elements for even deeper analysis. For example, if you are running a law firm, you might want to know which clients generated the most revenue last year, this year, and in the last quarter. This is a simple report anyone can pull from the billing system. You might want to go further and learn the level of profitability for each of those clients. This will require the application of some logic: you must first generate a list of the largest accounts and then do the math on the expenses against each account to calculate a profit margin. You might then want to know the percentage of your overall profits derived from the top 20 percent of clients. This requires first calculating who the top 20 percent are and then determining how much of the overall business they represent. The list of ways we can further slice this data is almost endless. The reality is that a good developer can give you any report you want if you just know how to ask and if the underlying data is there for the developer to report on.

For a step-by-step overview of the business intelligence process I implement with my clients, visit www.velocitymanifesto. com.

WEB-BASED DASHBOARD CREATION

An extension of the custom reporting we just covered is the creation of web-based dashboards that deliver key performance indicators in real time to the right people. Depending on the size of the operation, one dashboard could be enough, or it might take hundreds. Think of these dashboards as periscopes allowing a manager or an executive to have a quick view into the analytics that are truly important to the business. Once interviews have been completed to get leaders and other employees thinking about what reports they want and what rules and controls are needed, the concept of the dashboard becomes easy for them to understand. Every person in an organization could benefit from having a dashboard that delivers valuable statistics. I suspect that, in time, reviewing a dashboard of analytics will be a normal part of any worker's day.

The simple question to pose to kick off this process is "What information do you need to know in order to make good decisions today?" For some people there will only be a couple of metrics; for others there may be dozens. Once the important metrics have been identified, figure out whether the person needs drill-down capability to get beneath the analytics and see the underlying data. For example, I might want to see the sales numbers for today, yesterday, last week, last month, and for this day last year. I might also want all of this compared to the previous year so I can identify any trends. When I see an anomaly, I will want to be able to click on that particular number and have the system take me to another screen that gives me a breakdown of how that number was formed. So if I click on the sales for today because they seem extraordinarily high, I want to see a list of specific orders that were billed for the day. This is the drill-down, and it would show me if a salesperson closed a big deal that day or if the general volume of sales is growing rapidly.

It is easy to create a spreadsheet to use as a visual guide for what the dashboard would look like and what drill-downs will need to be created. Once completed, the spreadsheet is given to the IT people, who will then have to choose a tool for delivering the dashboard. Many tools are available as either hosted web-based applications or locally installed systems, so it is not difficult to build any configuration or appearance a user might want. Again, the only requirements are competent IT employees and the underlying data for them to reference as they build the dashboard.

RULES AND CONTROLS

Rules and controls can have a dramatic impact on the bottom line of any organization. The concept behind them is that boundaries can be set for the many processes, and when technology is used to enforce these boundaries, many undesirable circumstances can be avoided. We know the range something should be sold for. We know what the profit margins should be. We know whether we should lend money to a person with a certain credit profile. We know that if cash gets below a certain amount at a specific time of the month, we will have problems meeting the organization's needs. Many companies depend on several different team members to watch over all these boundaries and take action if any of them are breached. The problem is that throwing people at these conditions is expensive.

And the problem is not just the expense; humans are, of course, fallible and many times are not aware of breached boundaries until the problem becomes apparent in its extreme condition. Only when dramatic problems occur do we recognize what could have been seen had we been looking for it. This would not be a problem if we could hire large teams of people who never made mistakes. But everyone is trying to do more with fewer people to control costs and help grow the bottom line. This is where computers and software can play a tremendous role.

Let's assume we have built a good digital plumbing system and created a solid set of data that is easily turned into information and knowledge. With this powerful force behind us, we only need to add a layer of business intelligence on top to enforce boundaries and notify the proper people when they are breached. The truth of how our businesses operate resides deep down in our data, and this data can be leveraged to help us keep things on course if only we ask it to.

Here are a few examples of what I mean by rules and controls. Let's look at a medium-sized manufacturing/sales operation. It produces a large inventory of goods, about eight thousand different types of items. They ship all over the United States and have a large warehouse operation and a call center to take orders. A few specific problems are impacting this company's customer loyalty and the quality of their products. The first problem is that defective products are only noticed when someone in accounting or in the warehouse makes note of receiving return shipments of the same item over and over in the previous thirty days. Since those two positions are not tasked with solving the root problem of product defects, it might take twenty angry customers before the faulty products are pulled from the warehouse. Obviously, a software routine can be written that would send up a big electronic red flag if the company receives the same item back multiple times within a week. This information could then be sent directly to the quality assurance department so someone can fix the situation as soon as possible.

At the same company, the call center hustles and bustles with people on headsets taking orders all day long. The company's order-entry system can change the selling price of an item in order to complete a sale or offer a volume discount that the system does not calculate automatically. The problem is that some employees tend to be pretty liberal with their discounts when there are volume sales contests going on. They are measured on top-line revenue and thus do not really care whether the bottom line is impacted by their steep discounts to customers. Additionally, there is one person who has a friend call in from time to time, and on the buddy's purchases that day, all prices are dropped to $1 per item even though the average cost is $25. In other words, some employees are stealing from the company. With a rules-based business intelligence system, a rule can easily be written that will stop an order from processing through the system if it is sold below cost. Since there are legitimate times when products might need to be sold below cost, the rule can simply require a manager's sign-off before processing. This allows the manager to always be aware of this scenario so he can make a decision as to the appropriateness of selling below cost.

Both of these are prime examples of how software can enforce a boundary or control so that the right people are notified of a situation that needs attention *before* it becomes a major problem. Every organization has a frighteningly large number of boundaries that are administered by humans alone—frequently with no oversight at all. Sure, the accounting department at some point might catch a problem here and there, but by then it's usually too late.

Much like the reporting process described in the "Building a Technology Strategy" section of chapter 5, establishing rules and controls involves sitting down for a couple of hours with managers from every department in an organization and asking them a set of questions that will make them think about what boundaries could be developed to help safeguard the organization. Remember not to limit these questions to the technology currently available to the company; encourage employees to think outside the box as to what would be the perfect set of controls for their divisions. Normally, you can develop at least twenty to twenty-five new rules or controls that can be put in place, and it is not abnormal to end up with a list of over one hundred rules and controls dreamed up by employees from across the organization. When this list is generated, it's time for the IT department to figure out how to put them in place.

The implementation of these rules and controls will fall into one of three categories. First, you may have the capability to set the rule or control inside an existing application; perhaps it hasn't been done simply because no one has asked. The second situation is that you can create the rule or control by writing a bit of code that sits outside the application and monitors the raw data. When a condition is met or a boundary crossed, the prescribed action is taken. The third scenario is the most difficult: sometimes a new feature must be built into one of our applications. This process normally takes the longest to complete, and it has the most potential for impacting other services in the application.

What is most surprising to the executive teams is how easy many of these rules and controls are to put in place. This surprise is a good example of the gap that exists between executives and managers and the IT staff. Once someone mentions that the company could monitor a potentially negative condition with a two-hour IT investment, everyone throws their hands up and wonders why it took so long to get that safety measure in place.

Taken to its furthest positive extreme, an organization should have invested whatever time is necessary to create a complete safety net of sophisticated software to monitor every piece of an organization's data and analyze whether its goals are being met. Then a sophisticated set of rules for notification is built to alert the right people of issues as they arise. If this sounds like overkill to you, ask yourself why you invested in the software applications and their databases in the first place. Was it just to make accounting easier? Just to do things faster? Was it to keep up with the competition? Software can do all these things, but make sure you're using it to its maximum capability. If you really want to win in your market, setting rules and controls is a worthwhile process to invest resources in. The payback will come very quickly.

INFORMATION COMMUNICATION

Setting up the process we use to communicate information is the last step in making sure we use data well. Once the afore-mentioned processes have been completed, some thought needs to go into how you are going to share all of this new information with the rest of the company so that it can become institutional knowledge. One misconception needs to be eliminated right away, namely, that only certain employees should know certain information. Outside of payroll and HR information, nothing should be hidden from any employee. The more we share information, the more valuable it becomes. It is hard to predict who will derive value from different types of information, so there is simply no reason to guess who could use what information. As you hold the picture of fantastic digital plumbing in your mind, see the awesome stream of meaningful information flowing to all the team members, not just some of them.

The executives and the IT staff should come to an agreement as to which technology applications will be used to gather, hold, and distribute information now being created by many separate users. The goal is to create a company-wide knowledge base that can be referenced by anyone who needs it. I know that a few of you may be thinking it is dangerous to make all information available to everyone, since there might be people who leave and then take valuable inside information to a competitor. Point taken. You will have to choose whether you lose more from a competitor learning some internal metrics and altering their business to catch up or from restricting information from the people in your own organization. My belief is that we should make the vast majority of information available to everyone and limit its distribution only in very special cases.

The important thing to understand is that information is power—but only when as many people as possible turn that information into useful knowledge. The more each team member understands what makes the operation work, the better equipped they are to make decisions about their piece of it. This is not how most companies have been running things for the past several decades. Traditionally, the higher you are in the organizational hierarchy, the more access to information you have. And since knowledge is power, many people choose not to share it. But as the need for businesses to perform efficiently and profitably continues to grow, we will have to allow our digital plumbing to provide all the information possible to anyone in the organization and to release our fear of people taking information to a competitor.

Once processes like these are woven into the fabric of an organization, "aha moments" will abound as leaders begin to see results that are driven by the improved use of data. What a shame it is to have unused maps lying around—maps that could lead to buried gold—while executives ignore them in their struggle to put together winning strategies. It is certainly much easier to see where to go when the truth is gleaned from the data. It is also much easier to grow profits when the factors that drive the bottom line are thoroughly governed and observed.

Once you have assembled a fine-tuned digital plumbing platform and have woven processes into it that will allow it to grow and become progressively more efficient, you must focus on how you can leverage this new advantage. All the sophisticated technology in the world is a waste if you don't have a forward-thinking strategy. I suspect there are some naive leaders who think that with enough powerful technology on board, success is a given. This is clearly not the case, any more than having expensive golf clubs would guarantee shooting in the 70s on the golf course. You must have a clear picture of where your organization should be going in order to leverage the platform you have built to get there. Note that I did not say *how* you were going to get where you are going; I said *where* you are going.

I make this distinction because every organization has spent a lot of time thinking about its mission, vision, and goals. Its leaders can usually state clearly what the company is about, but the larger concern should be where it is headed. How will your company be relevant in the future? We are now seeing many large organizations that have invested lots of money in outdated technology infrastructures losing market share and simply fading away. In every case, they have failed to upgrade their strategy based on the changing demographics and dynamics of the marketplace. We seem to be ignoring that a critical component of leadership is being able to predict the future accurately in order to guide an organization toward a clear vision of the future. Wonderful technology infrastructure can do a lot, but it cannot save you from a business plan that has become stale.

For this reason, leaders today must have the skills and ability to keep the organizational strategy fresh and relevant. I run into many leaders who invest heavily in technology because it falsely makes them feel as if they are putting their operation on the leading edge. Understanding that technology is just a tool makes it clear that applying a world-class tool to an out-of-date business strategy will still lead to disaster.

On the other hand, a sophisticated infrastructure of digital plumbing, paired with a visionary strategy for being relevant in the marketplace, creates—you guessed it—velocity!

PART TWO

THE HIGH-BEAM STRATEGY

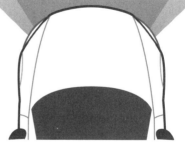

HIGH-BEAM LEADERSHIP

5- to 10-year view

- Visionary
- Willing to experiment
- Predictive analysis

LOW-BEAM LEADERSHIP

12-month view

- Monthly results
- Execution focused
- Lack of vision
- No future investments

CHAPTER 9

TRENDSPOTTING

*I look to the future because that's where I'm
going to spend the rest of my life.*
—George Burns

IN THESE TIMES OF HIGH-VELOCITY CHANGE, LEADERS
have to be able to see into the future to avoid miscalculations.
Just like high-speed drivers, today's leaders must use a "high-
beam" strategy to maximize their performance. The metaphor
of a car with its high beams on signifies the dynamics that occur
when a leader has an accurate view of the future: like the far-
ahead view provided by high beams, a clear picture of where
things are going over the next five to ten years in an organization gives a leader the ability to ensure that good decisions are
made in areas such as new product development, infrastructure
improvements, and staffing.

> **High-Beam Strategy:** Strategy that allows leaders to see the road ahead accurately and make sound decisions regarding their organization's future.

Feel intimidated? Well, if you think things are going to slow down and "return to normal," you are wrong. If you secretly hope that we will get back to a low-velocity world, you should put this book down because you'll be wasting time reading it. You need to look at the positive side of high velocity. I truthfully would be mortified if the velocity slowed because I like the pace fast—but then again I get bored easily. I also find that it delivers increased opportunities, and what fun would leadership be without an abundance of opportunities? Leaders, by definition, are supposed to be out in front of those that follow them. We are supposed to be the first ones on the trail, the ones chasing the best opportunities. That means we have to be able to look farther down the road than anyone else and understand how to take our teams along with us.

There have always been pioneers who were willing to bet on their vision of the future. The issue today is that the pace of the game has sped up so fast every businessperson needs to predict the future and recognize significant trends just to survive. Attempting to follow others or stick with the status quo is a fool's errand. Travel agents, the postal service, telecoms, TV networks, newspapers, radio stations, video stores, and book publishers—all of these businesses are struggling because they failed to plan for the future. The business leaders who are succeeding are the ones who have mastered the art of *trendspotting*.

Trendspotting: The ability to identify trends that will affect your organization or industry and then react to them by capitalizing on the opportunity or mitigating the risks they present.

For decades, leaders in the fast-changing "style" industries, such as fashion and entertainment, have relied on trendspotting for their successes. Spotting the next trend in skirt length or hairstyle could literally make or break these companies within a matter of months. Today, leaders in every industry need to be trendspotters because every industry is changing as quickly as, or even more quickly than, the fashion and entertainment industries are. Look at the field of law. Thirty years ago, if you wanted a will, a non-disclosure document, or any simple form agreement, you had to go to a lawyer's office and pay a fee. Today, I can use Google to search for a document and download a legal template in a heartbeat. It is not taking people very long to figure out they can do lightweight law on their own. The lawyers who spotted this trend and adapted their practice accordingly are the ones who are thriving. The others are either defunct or will soon go out of business.

Trendspotting is not easy. It's not like spotting a tsunami and preparing for the impact of one cataclysmic event. It is more like playing a cosmic chess match. Recognizing these variables and reacting to them appropriately is the intelligent move. Once trends are spotted, many organizations simply wait too long to adjust to the trends they anticipate.

It is not necessary to be on the bleeding edge in order to spot a trend and react quickly to it. Most industry-leading companies (like Southwest Airlines, Google, and General Electric) have the skill of staying current without taking on too much uncertainty. They do not take risks on something that has not proven its value, but once they see a trend that is here to stay, they push hard to monetize the opportunity it offers. This often means pushing employees, vendors, and partners to adopt these changes by a certain date. For example, Walmart saw that Radio Frequency Identification (RFID) tags—tiny antennas that can be used on any product to identify it from a distance—would be the next trend in managing inventory and pricing. Once it identified this trend, Walmart forced adoption of RFID through the supply chain so it could have an advantage over its rivals; it mandated that its top vendors use RFID on products by 2008. To date, their RFID mandate has not rocked the world as they thought it might, but they have used it successfully because they believe in the benefits the concept could provide.

Trendspotting is a skill and can be learned. Once it is learned, it needs to be turned into action. When I look back on my career so far, I am always thankful that I not only fought to understand how trends might play out but also betted on the trends I believed in: I was involved in selling PCs when the concept of personal computing was brand new. I built a company in the USSR a couple of years before it went through the revolution and became more business friendly. I started a webcasting company before webcasting was a widely known concept. In each case, I was building the company a few years before the trend it was built on became hot. I saw what looked like an obvious direction and was willing to take risks in order to take advantage of the trend.

Now that I have beaten you over the head with the importance of trendspotting, let's take a step back and talk about trends in more detail. The diagram below shows three categories of trends, working up from organizational trends through industry trends to the macro level. Every leader should have a clear picture of the most powerful trends in each of these layers:

Macro Trends
Globalization, technology adoption, generational change, workforce changes

Industry Trends
Regulations, demographic changes, new product, competitive pressures

Organizational Trends
Leadership change, product changes, changing values, tightening budgets

Organizational trends, or "local trends," are the trends that are very specific to an individual organization. These could include an upcoming change in leadership or a trend toward a more democratic method for making decisions rather than the CEO being a dictator. They could be trends that are exhibiting themselves in the town or state the organization markets in, or in new lines of business in which the organization is succeeding.

Industry trends are new movements in the fields the company works in. For a bank, these would be trends occurring across the financial institution markets, specifically in the banking space. They could be changes in the way consumers work with banks, changes in saving or spending patterns, or the consolidation of banks if they acquire each other. These are trends that are seen specifically within the industry you are in.

Macro trends are global movements. These are trends that impact everyone, including your organization. They could be economic, societal, weather related, or demographic. They are large changes in the world that you will either have to guard against or leverage for your company's growth; some may make it harder for your organization to achieve its goals, and others may represent great business opportunities.

Think of your organization as a boat making a journey across an ocean. While your boat is on the water, these three types of trends will impact your progress. At the macro level, there will be currents that pull you in various directions, and the weather will also affect you. The industry trend could be represented by the status of your boat. For instance, you would know that boats such as yours break down in consistent ways. Sails rip, oars break, and your model of boat is known to leak after twenty days on the open ocean; these are all trends that need to be monitored closely. Then there are the equivalent of local trends: say your crew is made up of older people who are not as physically able to make the journey as they once were, or that you have a new captain and the crew is divided on how much trust to put in him.

I actually just got seasick trying to use that metaphor. The bottom line is that your organization is like a ship on a very big ocean that you do not control, and if you do not have a pretty good idea of what might happen in the future on that journey, you may drown. Everything may appear to be fine as you leave the port under the shining sun, but if you factor in and prepare for the trends and conditions described above, you have a better chance of success than you would if you chose to believe the trip will be fine just because it seemed fine at that early moment.

Our organizations and strategies float on a shifting sea of trends. Some of these trends can be leveraged for gain, and some present huge risks. It's important for us to have a plan for dealing with both types of trend *before* we leave shore. Yet, some of the largest businesses I know have stumbled badly in the past because they had no anticipation of the trends that would eventually cost them dearly. IBM is a good example. Back in the late seventies, a handful of companies began pioneering the personal computer market. It was clear that PCs would soon replace mainframe-based terminals, and there was also a trend toward liberating users from the hegemony of the IT department.

With the gift of hindsight, we can look back and see that IBM badly mismanaged this trend and that it cost them billions of dollars. They have now abdicated the personal computing segment of the market to others. I have visited with many IBMers and heard all the excuses and reasons for why this happened. But let's be honest: IBM management ignored a powerful trend. It's that simple. IBM chose a strategy of not cannibalizing their mini-computer products, and in doing so they mistakenly gave away a huge product they could have leveraged in the future. They believed that because they were making strong profits selling their low-end mini-computer, they would be costing themselves money by pushing a possible substitute into the market. This was a conscious choice on their part. Sure, hindsight is twenty-twenty, but come on—it *was* pretty obvious even in the early eighties that PCs were going to be on every desktop in the future. IBM is doing fine these days, and they have done a good job of moving the company away from being hardware based and toward focusing on services and software. Even so, I have to wonder what would have happened had they used their R&D power and been visionary with driving the PC industry in the early days.

Trendspotting is all about gathering lots of information on the currents that might affect your organization, and then extrapolating how to leverage these trends to your advantage—or at least minimize the effects of trends that could kill you. The important thing is to give *conscious* thought to all these possibilities, including anything that will have an impact on what your company does. I emphasize the word *conscious* because trendspotting is a skill that is too important to be something that just kind of buzzes in the background. I am sure if we were to do a study of the reasons for the demise of the last fifty companies that fell out of the Fortune 500, we would find that the vast majority lost their momentum through a lack of visionary foresight as to where their markets were going to go.

Step back and be honest with yourself for a minute. Does your organization heed the basic wisdom of investing the needed time and energy in studying where the future might lead? Do you map out all the possible trends that could impact the organization and then construct responses that would leverage the changes, or at least mitigate the risks?

One has to wonder if leaders in the travel, music, and newspaper businesses spent any time thinking about the futures that faced them as they stood on the brink of major shifts in their industries. In each case, they seem to have been whipsawed by competitors from outside their ranks who came in and took major chunks of revenue. The first time you saw you could book your own flights, did you not sense that travel agents were in trouble? When you used Napster for the first time, did you not know in your heart that you wouldn't be buying CDs much longer? When free news content began to be delivered over the Internet in real time, and Craigslist began providing free ads, didn't it seem clear that printed newspapers would lose their place of importance in the world? So where were the leaders of these industries when interlopers like Steve Jobs took huge portions of their markets from them? Although I just picked on a few industries, I could easily have come up with a dozen others that have failed to see the same types of trends.

Trends can manifest themselves at varying speeds. At one end of the spectrum are the nearly instant changes. After the September 11, 2001 tragedy, airlines faced a massive and immediate drop in revenue that reordered their business and the businesses of the vendors that support them. Foodservice companies were impacted greatly when the airlines cut in-flight meals as a cost-saving measure. Then, when security was tightened and travelers could no longer carry on liquids in containers over a certain size, the consumer products industry had to begin supplying three-ounce bottles of many liquids frequently taken on planes. Many of these companies had very little time to calculate the impacts of the new security measures as they made the adjustments.

At the other end of the spectrum are trends that take a decade or more to manifest themselves completely. Many of these involve changes in the buying habits of the younger generations. An old-line brand like Levi's might experience long-term erosion in sales because a new generation grows up with a culture that is against anything their parents wore. In this case, the brand must construct a survival strategy that allows them to stay relevant to a generation that actually finds the brand distasteful because if its success with their parents. Another great example of this was Oldsmobile—your father's car.

You might believe that since these long-term trends move so slowly there is plenty of time to take action. But whenever you think you can take your time in reacting to a trend, forget it. Remember, the velocity of change is rapidly gathering speed thanks to technology and competition. The Internet has already single-handedly reordered entire industries in just a few years. I am not proposing, however, that all trends are based on technological advances; the men's clothing industry was rocked almost overnight when many companies went to business-casual dress codes. Social trends, economic trends, geographic trends, and new government regulations—many of which may have nothing to do with technology—can have dramatic impacts on companies.

So, have I convinced you yet of the need for spotting trends and reacting to them? I hope you aren't thinking: "This isn't a problem for me; I'm not that high on the ladder. There are people above me that have that responsibility." When new trends rock an industry, it's really not the very top leaders who get hammered financially—it's often the middle managers who find themselves out on the streets, looking at taking lower-paying jobs in another industry. In today's market, everyone—not just the CEO—needs to be trendspotting.

When I write about trendspotting, sometimes I feel like I'm stating the obvious, but then I open the *Wall Street Journal* and read another three stories of large organizations that have no clue how to place bets on the future. They spend all their energy mining the present for near-term profits and simply ignoring the future—all while they dance the death jig so analysts can see a small uptick in the stock price this quarter. Sadly, this strategy actually works for the CEOs of many major companies, since they can plunder their wonderful options packages, do a five-year stint, and then resign (or get fired) with a good chunk of money. Thankfully, there are leaders who actually plan on staying for the long term and who really do study the future and place wise bets. They are not always right in their predictions, but they are on target enough to sustain growth over the long term.

I spent my early career believing bigger meant smarter and large, multinational companies must have it all figured out. I reasoned that they certainly must have applied their resources to investigating any good idea in their market, and if they were not doing something, they must have tried it and given up because it didn't work. Today, I understand that many large organizations are living off ideas they had years ago. They actually become bound by their past success and have trouble developing new products or services—especially if they will cannibalize something they now make millions in revenue on. The bigger the company gets, the harder it becomes to imagine a future different from today, precisely because they have been so successful in the past. They focus on maximizing their present state—at the expense of their future.

In a small to medium-sized organization, the importance of understanding the tides of the trends flowing around you is a matter of survival. These companies do not have large capital reserves to lean on if they miss an important wave they could ride. The entrepreneurs who lead smaller organizations must navigate toward the future just to keep from being crushed by governments and large competitors. They must read the tea leaves well or they will perish quickly. And every once in a while, a clever entrepreneur spies a trend, bets everything on it, and wins big. Then we all sit around and marvel at how their big competitors missed that opportunity with one thousand times the R&D money and hundreds of employees dedicated to developing new products. This is the kind of thing that gets me up in the morning and makes me glad I was born in a free-market society—God bless America.

But it's not enough to be able to spot a trend. You have to know how to react to that trend. How is it that millions of people can view the same market conditions and understand emerging trends, but only a precious few see the diamonds buried within? Many people see upcoming shifts in their industry and simply do not have the will or the power to change the path of their organizations. In order to reach any goal, we have to have a pretty clear picture of the future conditions we may face. On top of that, we need a host of other skills as well. Does it take dedication? Hard work? Honesty, loyalty, innovation, a good team? Sure it does—it takes all of those. It takes every word that has ever showed up in the many bland and useless mission statements hanging on the walls of thousands of corporate lobbies at this moment. But before *any* of that matters, you must first have a clear and accurate vision of your future state. Without this picture, all the wonderful plans you worked so hard to get in place could be misdirected. The very real truth is that there are leaders all over the world who have built powerful organizations now headed straight for the graveyard because they are focused only on the present; designing a product that will be valuable in the future is very different from selling a product today. Recognizing that difference determines whether you are, as they say, the windshield or the bug.

Having the ability to do *accurate* trendspotting is going to be a mandatory leadership skill from now on. It is what will allow you to practice high-beam leadership. It is a talent to be practiced continually, and it requires a personal investment of time and brainpower. There is really no wizardry to trendspotting—it is a process that can be learned and refined. All you have to do is develop "vision precision," which we'll discuss in the next chapter.

CHAPTER 10

VISION PRECISION

*The very essence of leadership is that
you have to have a vision.*

—THEODORE HESBURGH

FOR MANY LEADERS, TRENDSPOTTING MAY BE intimidating. It is an essential skill, however, and it means that you will have to be a visionary for your company. This means you will have to be able to accurately predict the future. Stop for a moment and think about what I am saying, because this is too easy to just glide past. If you have the ability to look five years into the future, you will be able to spot trends and react effectively to them. With a clear and *accurate* picture of the future, your organization will not waste time going down blind alleys and can move quickly, knowing what needs to be done to reach its goals. You will be able to lead your organization in making investments that will pay off because they are based on accurate trendspotting and a solid vision of the future.

Ray Kurzweil is a brilliant futurist who specializes in mapping the progression of technology. He and his staff have developed many algorithms for predicting the growth and progress of various technology trends. What they have found is that many of the curves are quite smooth and predictable. Kurzweil's research shows that accurate trendspotting is possible, and he advises every leader to work to "see" the future, not just exist in the present.

Vision Precision: The ability to accurately spot those trends that will impact your organization. Emphasis on accurately!

So, how can you develop what I call "vision precision," the ability to spot trends? The first thing you have to do is to look at technology. Seeing trends in technological innovation is the key to understanding how your business will evolve. As a leader, you have to embrace and understand technology in order to be successful. I chat with thousands of leaders every year, and if I hear one more executive tell me she had to have her kids teach her how to use a cell phone, I am going to stick a pencil in my eye. Leading by example is the most powerful way to lead, and some of you need to pay attention to the example you're setting. If you want to be a powerful *modern* leader, then invest some time and energy in developing some *modern* skills. What kind of example of leadership is an executive who has to have an eleven-year-old teach her how to load a ring tone?

For the record, I have very technology-literate kids, and I work hard to stay ahead of them. It's not an easy task. No one learns how to use new tools by osmosis, and I work hard to keep current so that, if nothing else, my nineteen-year-old son will quit telling people he knows more about technology than me. When I build technology companies, I purposely do all my own technology work because I don't want the people who work for me to see a leader who promotes tech and yet cannot use it. Remember: powerful leaders lead by example—do you provide a strong model for leveraging technology? If not, no excuse is good enough. Don't tell me you are too old or your industry doesn't require you to use technology. Whatever business you are in, I am sure it is augmented by some kind of technology. One well-known banker in the state I live in did not answer an email from me, so I called his administrative assistant and asked what was up. She informed me he had more than fourteen hundred unanswered emails in his inbox and he really wasn't very good at email. *Really?* In the year 2010, you cannot even manage your email? I would guess he doesn't text message either. And he certainly has very little knowledge of the future trends in his industry.

To help you kick-start your ability to spot trends, I want to share with you some technology-based macro trends that are taking place now. These trends are commonly acknowledged, but they will continue to have a big impact on the business world. They are just a few examples—I won't elaborate on other trends based on globalization or generational changes. My goal in sharing them is to get you thinking like a high-beam leader. Millions of leaders around the world are aware of trends such as these but choose to ignore them—at their own peril. As you read through the list, think about how each entry will affect your industry and how you will react to it.

THE PAPERLESS OFFICE

A paperless office is one that has built computer-based applications for gathering information without the need for people to record the information on paper. A paperless office also avoids printing out documents as a means of information sharing and has the ability to complete transactions using only electronic work flows.

Long derided as a fantasy, this concept is now becoming reality in many industries. With the tools we have available today and the many benefits "going paperless" provides, it is just plain ignorant not to get there as fast as possible. This is an easy area in which to calculate return on investments, yet it is still a blind spot for many organizations. For the first decade after this trend first arose, we actually created more paper as we installed printers and dumped forests of reports into file cabinets. Like many trends, the paperless office has slowly crept up on us and is now maturing into a reality. For example, I have really enjoyed the fact that when I purchase something at the Apple store, they give me the option to receive my receipt through email. They have made a conscious effort to move all of their contacts and transactions onto a computer.

Think about the work flows in your operation today. How many are already using a completely paperless process, and how many still require paper to play a role? The first step in going paperless is the willingness to step back and review your business processes and assess where paper still plays a role. Once you can identify a handful of critical work flows that still involve paper, and have possibly identified conservative clients or customers still refusing to go electronic, commit to a plan to make the transition.

I know of one insurance company that convinced more than 50 percent of its doctors to file claims electronically. This change has allowed the company to grow by 18 percent in the last year without adding any employees. Along with this growth came an increase in profitability, which is no surprise. Their next step is to develop strong methods for enticing the rest of the doctors to file electronically, and to do so they'll be moving from the carrot to the stick.

At the industry level, it is wise to foresee how changes in passing documents electronically might impact you. As an example, think about all the changes that will occur in the mortgage industry when every home sale is closed using an electronic system. Because there are so many entities involved in supplying information at a real estate closing, the dynamics for all of them will change, as will the way in which the closing fees are structured.

SOFTWARE ARCHITECTURES

As we improve our ability to connect disparate software systems and create paperless offices, we create digital plumbing that re-engineers how entire supply chains function. The extension of our digital plumbing to customers and vendors so that data can flow seamlessly back and forth cuts back-office costs and increases the efficiency of information distribution. In order to connect our digital plumbing with the plumbing of the organizations we work with, there must be standards and connection methods. This is an area in which the software industry has made lots of progress in the last decade. We truly have made it easier to connect software systems together as if they were literally pieces of plumbing—fast, easily, and without leakage.

Let me explain a few aspects of connected software you should know about, and should be careful to oversee when your IT people are strategizing digital plumbing. The term *web services* refers to a set of standards programmers can use to easily connect different systems. By implementing web services, a layer of information is built into the software that allows a programmer from another organization to easily understand how to use and connect to a piece of code. This has been a huge step forward in how we design software. Frankly, it does not make much sense to buy or build a substantial piece of software without providing the ability for others to hook into it.

Web services have spawned a concept called *service-oriented architecture* (SOA). SOA describes the process whereby software architects build high-performing digital plumbing by dividing up functions into discrete software engines and then connecting and overlaying them to get the resulting informational view you want. When we overlay web services in order to create a completely new view of data, we get what is called a *mash-up*.

Service-Oriented Architecture: Technological architecture that focuses on connecting web services—standardized web-based applications—to be independent and interoperable, allowing many customizable data views and mash-ups.

If you're having trouble visualizing a mash-up, go to www.housingmaps.com and see how home rental information is overlaid on Google maps so people can easily find rentals in the area of a city they are interested in. This is possible because both Google and Craigslist offer their software as a web service and let any programmer use their data streams in any way they see fit.

The whole concept of using SOA to build mash-ups has opened a new world of online products. As of this writing, better than fifty thousand software applications have already been created that combine Google Earth's data with some other feed to create a unique set of information. This trend has huge implications for organizations and industries at large because it allows systems to connect with each other, aggregate data, and facilitate transactions in new ways, across a complicated set of disparate systems.

Web services and cloud computing have served to drive down the costs of IT support. As mentioned in chapter 2, *cloud computing* refers to the practice of renting computer storage and processing power from a supplier that has thousands of computers sitting in a building. Instead of buying your own servers and firewalls, you just rent them on an as-needed basis. You then have the capability to add more power or storage simply by making a phone call. By renting/leasing IT services, organizations no longer have to build their own data centers and fill them with servers. This frees up IT staff to function as software analysts rather than software builders and janitors. These trends, because they are web-based, also free up employees to work from any location, since they can access organizational tools from any machine with a web browser.

Knowing that these new paradigms are going to allow the connection of plumbing to happen faster, more securely, and more simply, how does the world look different? Will you invest in getting there faster? Will your business be impacted negatively if your industry goes completely electronic with transactions and document flows? Will the product or service you provide even be needed in the future? If you have a hard time making this thought real, let me put it this way: Would you want to be a bike messenger or document delivery service in the legal industry if all communications between lawyers and courts go electronic? Don't look at the world through rose-colored glasses on this one. Digital plumbing coming together across supply chains is going to radically shift how your organization operates. It is better to understand that and diversify than to hang on too long and perish.

CROWDSOURCING

The process of using a website to coordinate the electronic herd into a specific activity is called *crowdsourcing*. The reach and power of the Internet is now allowing organizations to enlist millions of users and customers for help carrying out tasks, and thus activities that used to cost lots of money can now be accomplished for pennies. These include things like R&D, market research, information generation, and tagging.

There are many examples of organizations using crowd-sourcing to build huge new properties (www.wikipedia.com), provide free services (answers.yahoo.com), and guide product development (www.dellideastorm.com).

You can also find a full list of crowdsourcing sites, broken down by category, at www.velocitymanifesto.com.

This trend is being driven by the reality that millions of new "netizens" access the Internet each week, and many of them have extra time on their hands, which they will devote to something they see as useful and/or fun.

Knowing that you have access to this free, electronic work crew at any time, think of activities you could outsource to them. Is it hard for you to think of any? Recently, there was a young developer at Google who built a game that could be used to index photos in an extremely accurate way. He flashed photos to two different players and gave them only seconds to type in all the words that would describe the photo. The players got the enjoyment of competing with someone else to see who could think faster, and Google got thousands of photos indexed free of charge. The players in some cases had no idea they were being crowdsourced.

Crowdsourcing has come about because we now have the ability to distribute work very easily across the world to millions of people for very little cost. And the base of potential users to participate in crowdsourcing projects is expanding, especially when you consider China, India, and Africa getting online. By now, you should get the picture that some work you currently pay for could be done for nothing, or at least very cheaply.

THE METAVERSE

One step further up from crowdsourcing on the ladder of the future is the concept of synthetic worlds. Science fiction writers have long written about alternative electronic worlds we could "live" in—either by visiting them or by uploading our brains to them. We are still a ways off from uploading our brain anywhere, but we do have online worlds today. The first commercially viable one seems to be Second Life, which has crossed over from an online world visited mainly by hard-core geeks to a place where real business is done on a daily basis. Second Life contains real estate and a monetary system, and real people are represented in this virtual space by avatars. It is a world unto itself, enforcing its own rules, even for things as basic as gravity. While it may at first glance look like a video game world, there is no set game to play—it's a place for people to live and interact.

Linden Labs, the creator of Second Life, recently added an enterprise version of the platform that companies can buy; it gives them their own private, firewall-protected virtual world in which to hold meetings and collaborate. For some organizations, building a presence in online worlds like Second Life makes a lot of sense. This very new trend will undoubtedly blossom over the coming decade. The cost to experiment in virtual worlds is low, and the returns could be phenomenal for early players.

So here is the hard part for most people. When you have not played around in Second Life or one of its competitors and thus do not have enough vision to understand why taking part in a synthetic world makes sense for commercial reasons, it is impossible to imagine how it can be of use to your organization. You may be thinking this is so far out it's a waste of time to contemplate, but look at it as a test of your creativity. This is the kind of trend that seems crazy now, but it will make loads of sense in ten years—much like how it would have been crazy to think we would have something like the Internet from the perspective of 1992. We had the Internet then, and some visionaries were telling us it would be huge, but it was too hard to imagine. To avoid repeating past mistakes, it is a good idea to gain the skill of accepting that a trend will be powerful, and also understanding that it might be a decade before it is commercially viable.

IBM should get credit for being an early user of Second Life and developing stores and businesses there for clients. IBM now holds international meetings in Second Life at an online facility they built there. Imagine how much easier and cheaper it is to send an invitation to twelve people from around the world and have them meet at your "office" online. Second Life even allows them to do immediate speech translation, so they can speak to each other and have the system translate what they are saying as if they were at the UN.

An intriguing thought about meetings in synthetic worlds is that they do not need to imitate the real world. Because they happen in a virtual world, participants can use visual cues to give a whole different spin to the meetings they hold there. There is a designer who has developed a new concept for online meetings: the floor of the meeting space has the word *Agree* on one side and *Disagree* on the other. As topics come up, attendees can walk over to the side of the floor that signals their position on the topic. As they stand there, a graphical tube grows up around them and shows how long they have held that position. If they change positions, the tube bends and shows that the attendee changed his or her mind. There are other unique components to this designer's idea, and it is a fantastic example of how activities in a virtual world can actually improve on the real world.

With trends such as this, it will take a competitor leading the way or customers demanding that a company use it before many leaders will "get it." Knowing how to leverage trends starts with identifying and understanding trends. Then, leaders must force themselves to create a picture of a future that is different from today and extrapolate how that future might be a positive or a negative for their business.

Just for fun, let me point out a few non-technology trends that are looming on the horizon. None of these should be shocking to you, yet I *promise* that many leaders who will be substantially impacted by them have no plan for leveraging or defending against the changes these trends will bring.

1. According to the U.S. Department of Labor, the average worker coming on the market today will have had an average of ten to fourteen different jobs by the age of thirty-eight. This means that most people will have completely abandoned the notion of staying with a company for their entire career. Employees will stay at their jobs only as long as they feel fulfilled, feel they are making a difference, and feel there is a financial incentive to hang around. Fail to provide any of the above, and it will be only a matter of time till they move on. Not only that, it will be the A players who move on the fastest. They will be welcomed with open arms anywhere they interview. Your C players will stay with you longer because most others do not want them. The B players will split down the middle: the half that is more upwardly mobile will want to move on; the half that are content to stay in one place will hang on with you. What will you do to improve your organizational culture to keep the A players?

2. People are living longer. In fact, some experts are saying that the number of people who live to be more than eighty years old will skyrocket in upcoming years. With the baby boom generation beginning to retire, we could see approximately 35 million people leaving the workforce in the next decade. What will this do to the various businesses involved in hiring and recruiting? What

will it do to employee-intensive businesses? What will it do to our willingness to allow more immigration?

3. Globalization is reordering which countries are best suited for different parts of the supply chain. In the United States, we are very good at R&D and innovation, and we have a stellar system for growing entrepreneurs. Yet China, Germany, and Japan have distinct advantages on us when it comes to manufacturing. China and India, among others, have wage scales for labor that make them much more able to handle basic tasks cost-effectively. In time, we may see countries move very consciously to specializing in certain areas of the supply chain for goods and services.

4. Governments are inexorably moving toward regulating more and more of our lives and businesses. With every negative event or trend that happens to humanity, a government makes its best attempt to play the role of protector by enacting laws that can prevent these issues. Yet the line between protecting and restricting is a fine one. Overregulate businesses and you will take away entrepreneurs' desire to start something new, and you will create a situation in which prospering economically becomes so difficult some people will just quit the game altogether.

Many people look back with the advantage of hindsight and become frustrated when they see where they might have missed an opportunity. It is even easier to point out where others have overlooked an important trend. But looking backward is pretty worthless. If you are really going to be a strong leader, you must improve your ability to see where the winds of change are blowing and get your sail up.

One of the benefits of having been raised in the technology industry is that I had to learn at an early age the value of seeing ahead. Seeing ahead meant not only being able to spot future trends but also being able to estimate the timing of when those trends would make a market impact. It is almost useless for a person to have a great picture of technological trends going forward if they cannot also foresee fairly accurately when the changes will come. Many futurists think it is enough to predict interesting things that will happen in the future, but knowing both *what* will happen and *when* it will become capable of being monetized is the truly impressive skill.

I learned this lesson early on when I made financial bets on changes in the market. I misunderstood the speed at which the public would grasp these technological changes and found out we were too early in the market and did not have the patience or staying power to wait for our technologies to become mainstream. The result was that I was applauded as a visionary but was unable to turn my vision into the wealth that would have been possible had I had a better sense of *when* that vision would come to fruition.

This is where the concept of vision precision comes into play. Once you have both accurately predicted what changes will occur and when they will come to pass, you will be able to map a strategy to leverage this future. Businesses in the United States have not been very good at this recently, as they have become short-term thinkers struggling for short-term results. The Japanese have long had a more effective model because they actually can conceive of a hundred-year plan. Most U.S.

firms are lucky to have a plan for three years. A wiser approach would be to build a hybrid of the two philosophies in which there are one-year, three-year, ten-year, and twenty-year plans. The longer-term plans would be tightly tied to the ability to accurately project a future state of the industry and make moves today that are at least on a path that makes sense.

In order to build the skill of seeing accurately into the future, we have to change the way we currently think, and we also need to dedicate time and resources to the pursuit. Check out your calendar. When was the last time you spent three straight hours brainstorming what the future will look like? Did you write down your ideas so you could check years later to see how accurate you were? When did you last use a formal process to build a picture of the future and then link your current strategy to getting there? If you haven't done any of the above, don't feel bad; you are not alone. Ask yourself, however, about the ROI that could come from building this type of skill. Is it worth a little investment of your time and energy today?

CHAPTER 11

THE FOUR QUESTIONS THAT CAN SHAPE YOUR VISION PRECISION

Discovery consists of seeing what everybody has seen and thinking what nobody has thought.

—Albert von Szent-Gyorgyi

WITH VISION PRECISION, YOU WILL BE ABLE TO START spotting trends and when they will impact your organization. The more confident you are in the accuracy of your vision of the future, the more prepared you will be to lead your organization. You will not be bogged down by nagging doubts. You will be able to keep your eyes on the prize and move forward with great velocity.

I will share in this chapter some of the questions that I ask myself when trying to glean where the world is moving. In order to develop an accurate view of the future you can believe in, you and your team must consider information from a variety of sources that can help you triangulate what you see as the truth. It helps to develop what I call your "river of information" to create fertile ground for germinating a clear vision. Your river of information must include the insights of many individuals, not just one person (especially not just your own).

Science fiction writers, specifically cyberpunk writers, fascinate me; they have an amazing ability to paint what are often very accurate pictures of what the future holds. I read a decent amount of science fiction because I find that their landscapes add to my view of the future in critical ways, and that their visions often come true. For example, I love the concept of man's integration with machines, as described in William Gibson's 1984 book *Neuromancer*. The concepts he wove into the story are amazingly like what we see happening today.

So, back to the title of this chapter: Why is asking questions an important tactic in creating your future vision? Because they will force you to consider the world from a specific viewpoint. For example, if I ask you why it is that the opposite sex seems to view the world through a different lens, you are forced to put yourself in the opposite sex's shoes in order to analyze why. By asking intriguing questions about where the future might go, you have the potential to develop solid opinions on the topic through your answers. Questions make us think for ourselves. Questions inspire us to consider new possibilities.

It is impossible to build an accurate picture of the future solely by using your own imagination. It requires having thousands of data points, and the quality of those data points will be tied directly to the quality of the stream of information available to you. With a river of information coming from many different sources, you and your leadership team can begin to consider four important questions that will help refine your vision precision:

1. WHAT IS THE BEST, MOST EFFICIENT WAY TO PERFORM CRITICAL TASKS IN YOUR INDUSTRY?

Set aside what is possible today. Set aside the ways you have seen it done for decades. What would be the very best possible scenario? This question is much bigger than it first appears, and it is exceedingly difficult to answer because it requires us to see past our current view of our organizations. Let me give you an example. If one was to ask two hundred years ago, "How can people get from place to place in the fastest manner?" the answer would have been, "By horse and buggy." To fully answer the question from the perspective of two centuries ago, you would have to open your mind up and see that the future might bring a horseless, self-running wagon. To be even more visionary, you would have to be able to foresee that we would also figure out a way to fly.

Let's look at today. We now have the ability to fly to outer space, and thousands of people transverse the oceans on aircraft daily. Although these were big steps, the most efficient way to get from point to point would be to instantly travel to a desired point. Science fiction has been describing this for years. *Star Trek* had the transporter for short distances, and many writers talk about using wormholes to go longer distances. Do you really believe this will happen, and if so, when?

Here is the bottom line on figuring out the absolute best way possible to do things: if we can imagine a way to do something that is better than today, we will eventually figure out how to do it. So I believe we will one day figure out how to transport ourselves from point to point instantly. The pivotal question is when.

Now let's apply this question to a business scenario. If we would have asked ourselves back in the eighties what would be a better way to communicate with people around the world, or even across town, we might have been able to come up with a couple of ideas. We first would have said we need a free or almost free method. We would also need a unique address for everyone we want to talk to. We would like to have the choice of talking to them or writing to them. We would also like to know if they are available to chat right now, or if we should just leave a message. All of this would have seemed a bit far-fetched in 1987. I know this because I started a company in the Soviet Union at that time, and we had to either write letters—which were delivered on an unpredictable time schedule if they were delivered at all—or try to make a call on the extremely unreliable Soviet phone system. There really was no way for us to communicate meaningfully without speaking to someone in person.

Today, if I started a company in a foreign country, I could talk over Skype free of charge, instant message in real time, and use email without a time lag. I could communicate with almost anyone as if they were only a mile away. So let's step back and think about what it must have been like to be a telecommunications company in 1987. Were they forward-thinking enough to understand that we would develop such means of communications to solve the frustrations of our limited technology? Did they make moves to leverage these trends or mitigate the dangers? For the most part, they did not. Had they done so, they would have purchased Skype rather than letting eBay have the pleasure.

We have an *amazing* ability as a species to solve problems and make improvements. One way to enhance your vision precision is to understand that we *will* solve the issues in all industries. We will find efficiencies and knock out things that don't work the way we would like them to. So all you have to do is look at the industry you are in and identify all the things are that are wrong with it. What is inefficient within the supply chain? What information does not flow well? List the top three things that cause friction in your industry, and I promise you they will be solved one day. The only question is when. Next, you must begin to study how the solution to that problem will impact you, and find out whether you have any ability to solve it on your own.

Just to prime the pump, here are some examples from different industries:

- Banks used to be open only when it was convenient for them, not when it was convenient for families in which both adults work. Now, many banks are open on Saturdays, and a few renegades even open on Sundays.

- Walmart combined two types of stores (grocery and mercantile) into a megastore so time-crunched customers could combine two stops into one. Then the Internet came along and Walmart began offering to deliver items without requiring customers to visit the store at all.

- Dentist visits used to be torture, so many dentists started making their practices more like spas. Now it's not quite as bad to go get your teeth cleaned.

- Companies that recognized cooking in an oven just takes too long began selling the microwave oven.

All problems beg for solutions, and all of the current solutions are still evolving. As we find new solutions, there will be unintended consequences to businesses and society in general. The more you learn to understand that we can fix problems and frustrations, and the more you see your market in these terms, the better you will be able to predict what will happen in the future.

2. HOW ARE THE TRENDS YOU HAVE IDENTIFIED GOING TO MATURE OVER TIME, AND WHAT WILL THEIR IMPACTS BE?

As you work through the high-beam process, you will build a wonderful picture of the shifting ground you are now dealing with in your industry. You will likely see that many of the trends are simply steps toward solving some frustration or friction point you deal with every day in your work flows or processes.

This second question is all about extrapolating out the effects of the three types of trends—organizational, industry, and macro—we discussed in chapter 9. Each type will have a specific impact on your future. If you make a substantial mistake in your predictions and apply too many resources to a trend that doesn't pan out, you could waste lots of time and money. If, however, your prediction is right, you will have a huge advantage in the marketplace.

The surprising thing to me is how little trendspotting is practiced by some really smart people. Of the four questions we are going to discuss, figuring out how to gauge the impact of the trends you see coming should be the most obvious. It should be easy to extrapolate a known trend to its conclusion and to make a reasonable guess as to when it will come to pass.

In my consulting work I see some very smart C-level executives who simply do not invest energy in this area. They are smart enough to see trends, but they do not spend much time thinking about how the trends will play out. In some cases they have given a little thought to the ultimate conclusion of the trend but have not made a timeline of its stages and how each one will impact the company. Their flaw is that they are overly focused on executing today's plan. They mistakenly believe that if they pound out wonderful execution today, tomorrow, and next week, everything else will work itself out. The problem is that they are executing a plan that will in some cases become off target in a matter of months. They don't understand the importance of trendspotting and trend leveraging.

Let's consider the trend of consumers wanting healthier food choices from fast food restaurants and play it all the way out. In the United States, people want healthier food choices from fast food restaurants because they want to lose weight and do not have time to prepare healthy meals. Will people continue to want to lose weight? Probably. People do not like to exercise, but they do love to eat, so they will probably continue to be overweight. Will people want to spend more time preparing healthy meals in the future? Probably not. People have been trying to spend less time in household food preparation for centuries.

So, given the trends that consumers will probably continue to want to lose weight and will probably not want to spend more time in food preparation, the demand for healthy fast food will probably increase. We can observe how the healthy-eating trend is playing out right now. Healthy fast food chains are doing well; unhealthy chains have had to add salads and other healthy fare; and grocery stores are offering ready-to-eat healthy dinners people can pick up and take home.

We will soon see partnerships between companies like Weight Watchers and McDonald's or Jenny Craig and Burger King. If not this, we will see a Weight Watchers fast food chain where you can choose your meals by calorie quantity. At the drive-through you will say something like, "I want the 400-calorie chicken lunch, please." What are some other possible impacts of this healthy fast food trend? Drug companies are developing pills that block the effects of bad foods so we can still eat them without the weight gain. The medical profession will continue to develop surgical procedures to assist in weight loss, including gene therapy to help those who are genetically predisposed to obesity. And the list goes on.

3. WHAT OTHER INDUSTRIES ARE AHEAD OF YOURS, AND WHAT CAN YOU LEARN FROM THEM?

As a professional speaker, I have the opportunity to work with people in many different industries. I get to see which industries really push the limits of forward thinking and which are mired in age-old habits and plagued by inertia. I have run into a few people lately who make a habit of visiting trade shows from industries that are quite different from their own, just so they can observe what others are doing and learn how others' new ideas could apply to their businesses.

Some people might ask, "What other industry should I look at?" You shouldn't think too hard about trying to find one that is close to yours. That is not the point. The point is to study how any other industry works. How do they communicate? How electronic are they? How do they work with clients or customers? How do they solve problems? Do they have an association, and how does that group help the industry? There may be hundreds of aspects in which the industry is like yours and another hundred in which it is different. It is good to observe the differences and digest how they are solving problems and getting things done in unique ways.

If you are in a non-technology-based industry, you might want to attend a technology conference. By virtue of the fact that the dynamics of the technology world are faster, more competitive, and on the cutting edge, you tend to see more innovations in processes and corporate culture. Because they develop and sell technology, they are often better at using it internally than is the average enterprise. There are good lessons for any business in the success of companies like Apple, Google, Microsoft, and Cisco. All have used innovative concepts in standards, design, human resources, and corporate culture to great success.

Speaking of Microsoft, they have been known to use the phrase "embrace and extend" as a motto. Observers have been split on whether to view this in a negative light—some translate it to, "Steal others' ideas and then add our own proprietary components to lock out competitors"—or with a more positive perspective—embrace current standards and then extend them with our own special features. I love the concept of embracing the good work and standards that have gone before, then adding my own "special sauce," so to speak. It is all about standing on the shoulders of those who have gone before you. It also relates to this third question we're using to develop your vision precision. What systems, processes, or innovations can you embrace from other industries and then extend into your world? Even if you run a flower shop, you can pick up improvements from looking at a local car parts store. In a world moving as fast as ours is, it is good to be able to take a few shortcuts by observing the successes of others and making them your own.

On the flip side, if you are in the technology industry, attending a manufacturing conference would give you a totally different view on doing business. The issues manufacturers face are completely different from those technology companies face, and therefore the conversations are very different. You might see breakout sessions on lean manufacturing and hear speeches on import tariffs. There would be discussions about safety programs and the latest thinking on productivity increases through the use of robotics.

A sharp leader of a software company who decided to attend a manufacturing trade show might latch on to the lean manufacturing concept and find ways to apply it to software development. The company may then go on to experiment with code generation systems as an analogy to robotics, since both can eliminate an organization's dependence on human hours devoted to manual tasks.

When I was a CEO out building my technology companies, I never realized how limited my thinking was until I became a consultant and speaker. When you go to the same office and are surrounded by the same people every day, and you are thinking about solving the same problems over and over again, you invariably become a bit myopic. Only by getting away from your usual world and observing other, very different environments can you build the skill of assimilating ideas from other industries into your own. Perspective is a powerful thing, and looking at the world through the eyes of another industry is a great move.

Now, on to the last question . . .

4. WHAT IS THE TIME HORIZON FOR THE MOST IMPORTANT CHANGES YOU SEE COMING IN THE FUTURE?

As I have already discussed, timing is crucial. Many people can speculate accurately on what will happen *someday*. For example, futurists will say things like, "*Someday* we will not need glasses anymore because eye surgery will be so advanced that anyone with vision problems will get it." I believe this to be true. My wife and many of my friends have had Lasik surgery and are doing great. The real trick is not being able to make this general prediction; it is knowing what the adoption curve will look like over the next few years or decades so you can predict when the local eyeglass store will go out of business or whether the market is shrinking too quickly to pursue a career in optometry.

Changes in the market rarely happen in a matter if months—they normally take years. Visionary people tend to get overly optimistic about how fast adoption will take place; people with no vision tend to be pessimistic and wait too long, convinced that a trend is not really happening because not everyone is adopting it.

It is dangerous in the extreme to become a predictor of the future and not be realistic about the timing of the changes. Make a prediction and begin to invest resources too early, and you may waste them before you get an ROI. Wait too long, and you will get behind your competition and risk never catching up. The timing is *very* hard to gauge. I have been too early much more than I have been too late when it comes to adopting new trends, and being too early has taught me that the bleeding edge is just as dangerous as moving too slowly. In fact, the thing you want to be these days is a "fast follower." The concept here is that you have a better chance of making a good investment in a trend or idea if you are not the very first—but also not the fifth, sixth, or seventh—into the market. Instead, you want to quickly follow the first movers who are trying to capitalize on a blossoming trend.

Be careful not to think that you can be a fast follower just by watching what others do and waiting for the right moment to jump in. A true fast follower has had a trend on their mind for a while and has been ready to move on the opportunity. You have to have understood a trend, watched a first mover navigate it, and identified their mistakes—then you can follow their lead successfully.

This does not mean there is never a time to be a first mover. If the market is ready for your move and it's clear that you can get a return on your investment, by all means, move. There can be as many plusses as minuses in being a first mover. Someone has to go first, and there are plenty of success stories of leaders who were first out of the gate.

As you see, establishing a time horizon for when you will implement trends into your organization is a key part of the skill of trendspotting. A good example of a trend that needs to be prepared for is the retirement and old age of the baby boomers. The baby boom generation has had a big impact on buying trends, housing markets, employment statistics, and more. Many predictions are being made as to how this population bulge is going to impact the country for the next couple of decades. One aspect futurists mention is the growth of assisted living centers, nursing homes, and senior care services. This is a pretty obvious conclusion based on the current reality that the infrastructure for taking care of the elderly is not big enough to handle future demand. In addition, as medical care improves, people live longer. I doubt anyone would argue against the validity of this prediction. If you are building assisted living centers, the trick is to understand exactly when to build and at what size. If you build too early, you will be unprofitable until the site will fill up. If you build too late, you may have to deal with competitors who have locked up the market. Being even five years off in your projections could spell disaster.

By pondering these four questions and studying your answers, your team can sharpen its trendspotting skills and its vision precision. Remember, this is a dynamic process. I've been doing this for years and now enjoy reading about changing statistics and new inventions. With each one, I get the chance to confirm or alter my view of the future. I also get to adjust the timetable I have in my head as to when the changes will come and what the adoption curves might look like. The longer I do this, the better I feel about the accuracy of my picture of the future.

In the next chapter, we'll look at a specific trend: white collar lean, a concept that is becoming a powerful force for lowering back-office costs. Let's look at this trend in detail and see how you might want to apply it in the near future.

CHAPTER 12

TRENDSPOTTING CASE STUDY: WHITE COLLAR LEAN

Life is full of obstacle illusions.

—Grant Frazier

AS I PROMISED IN CHAPTER 4, I WANT TO GO INTO some detail on how an organization can institute a "white collar lean" process in order to improve productivity. White collar lean is a great example of a trend that can be universally helpful for any organization. It is not a bleeding edge trend, and many people would think of it by another name: business process management. I love this concept because it leverages technology to provide both operational velocity and profit velocity.

> **White Collar Lean:** The concept of streamlining office processes to reduce waste that comes from repetitive or unneeded actions, delays in decision making, or anything else that consumes time or resources unnecessarily.

In every industry, companies want to make back-office operations more efficient and speed up the performance of white-collar tasks. By the way, this is a great area in which to apply the digital plumbing you have worked so hard to build. What makes your plumbing valuable is learning how to apply it to either raise top-line revenue or lower back-office costs. With white collar lean, we are going to attack the back-office costs in order to discover a few percentage points of profit that may have been lying on the ground.

This should be obvious, but just so I have said it: in order to create performance velocity, leaders need to refine many of the basic, labor-intensive processes all organizations have. This can be hard to do because most companies have done these work flows and business processes the same way for many years, and thinking of new ways to do them is difficult. This is why people are often brought in from the outside to guide the reconstruction of the manufacturing flow. If you run an organization that is primarily a white-collar operation, pay close attention to this section of the book. It could yield powerful results and dramatically speed up business processes while also lowering head count.

A bit of background on the "lean" movement might be helpful here. Lean manufacturing is a paradigm that was developed about thirty years ago in Japan. I won't spend pages and pages explaining the process, but if you want to know more, you can find much more information on the Internet. We'll just cover the important points for our purposes.

There is a lot of wasted motion in everyday tasks, and a lot of this waste is invisible to us. It adds up over many repetitions and across many employees. If we can lower the waste, we improve productivity. The process of "leaning out" something is simply identifying the wasted motions or resources and finding ways to stop the waste. The waste could be in time, resources, or materials. In the back office, the waste normally comes from duplicated efforts, lack of visibility into the status of a task, or doing things by hand that could be done by computer. Reconfiguring a work flow in completely new ways can often mitigate waste. Of course, the tough thing about this is breaking years' worth of habitual behavior.

Before you can improve a work flow, you have to identify what your goals are and what steps in processes are necessary (and unnecessary). For some strange reason, steps that are not really needed anymore often stay in place because we have been doing them for years and have never really thought about taking them out.

Technology can be applied to many back-office tasks in order to lower the dependence on human beings and move the work to computers and software. In theory, computers never call in sick, and they never have a bad day. Once they are programmed to do something a certain way, they never deviate. We now have the ability to program software that will make the same decisions a human would make given the same set of data.

Here are some examples:

- The antilock braking system on your car makes decisions about the condition of roads and automatically alters the braking process for you.

- Insurance companies have adjudication systems that process electronic claims and, with no human intervention, make the decision on paying them.

- Banks are using "loan-decisioning systems" to say yes or no to granting loans.

- Your iPhone studies the mistakes you make when you text and adjusts itself so that common mistakes are corrected automatically.

Trust me when I tell you that we have all the technology we need—off the shelf—to take huge costs out of the back office. To apply it, we just need leaders with vision for what technology can do.

The following steps are the ones I use when helping organizations build a perpetual process for practicing white collar lean. Note the word *perpetual*—once this process is in place, it should never cease. There will always be new ways to apply the process as an organization grows, develops new products, or is impacted by market conditions.

STEP ONE: THE NECESSARY TOOLS

Before you can actually use technology to make a process lean, you have to have the technology, so let's talk about tools. You are going to need a reporting system that can be customized, and you will need the ability to write custom code on top of your database. This is because you will have to reach into the data and monitor conditions.

You will also need an ad hoc work flow system so that you can rebuild the processes you do by hand into a series of digitally controlled steps. An ad hoc work flow system is a software-based platform that allows you to create any type of work flow you can imagine; it gives you the tools to create your own queues and rules. These queues and rules are simply decision points and the procedures that govern those steps in the work flow you create. An ad hoc work flow system is much like a spreadsheet. When you first get Excel, it does nothing valuable out of the box. You actually have to build formulas and fields into it before it can do anything, but all the tools to build anything you can imagine are provided. A work flow platform gives you a collection of tools that help quickly assemble a digitally enhanced version of a business process you now do by hand and on paper.

Ad Hoc Work Flow System: A software-based platform that allows the user to create a series of queues and rules to govern the components and steps of a work flow.

You may need more tools than just these, but they are the basics to get you started down the road.

STEP TWO: KNOWLEDGEABLE TECHNOLOGISTS

You will need IT staff, or a set of IT contractors, who know how to use the tools listed above, and there should preferably be someone on that team who has a good understanding of the business and good relationships with the people in the office. This will help because this IT person will have less of a learning curve in understanding the processes that need to be improved. In the best-case scenario, there would be a software developer dedicated to the white-collar-lean process for years at a time. This lets them become very familiar with the tools you are using and the expectations you have for success.

STEP THREE: CHOOSING YOUR EARLY TARGETS

Identify the three most important work flows in your organization, and then the three most wasteful work flows. Sometimes these are the same work flows, and often they are different. A work flow or business process simply consists of the steps it takes to complete a discrete task in the organization. These tend to be things like taking orders, shipping products, engaging clients, and taking new members on board. They can also be processes like terminating employees or dealing with angry customers.

STEP FOUR: PICKING THE FIRST PROJECT

Once you have a list of the most important and most wasteful work flows in the organization, you are ready to identify one that you will work through and improve. The decision as to which work flow you will do first needs to be guided by which would be a good testing and learning example that could serve as a model for all to follow. You will likely find the first few work flows you take through the process of moving from a "by hand," people-intensive operation to a highly technology-augmented operation to be difficult. The more your team gains experience with work flow tools and the mapping of new electronic work flows, the easier all the rest you must do will become. Sometimes the answer to which work flow should go first is driven by what needs to be improved most urgently. At other times it might make sense to pick a smaller process—even if it's wasteful—to serve merely as a good test. In either case, pick just one for now.

STEP FIVE: THE FIRST MEETING—DRAWING THE CURRENT STATE

Once the work flow candidate is identified, call a meeting of the people who are involved in handling the process and bring them into a room with a whiteboard. The IT staff will also need to be there to help you identify where technology can be used to eliminate waste. It takes a creative, almost artistic mind to see how a process can be improved by reaching into a palette of technology tools. Map out the steps of the current process on the whiteboard so that everyone can see how it is done today. Capture this information on paper or in an electronic file so you have a record of it. It might even help to use color coding and note problem areas in red.

STEP SIX: MAPPING OUT THE FUTURE STATE

On another board, or below the current diagram, draw a new diagram that shows how the process can be augmented and/or streamlined by using technology. There may be specific areas in which technology can improve unsophisticated processes, such as digitizing data so that it is accurate, normalized, and entered only once. Work flow platforms will let us write rules to govern the flow of any process so humans don't need to route or monitor tasks. Automated triggers can be written to take action based on whatever conditions we set. Alerts can be created for tasks that go overdue, and tasks can be automatically routed based on a variety of conditions.

STEP SEVEN: BUILD THE NEW ELECTRONIC WORK FLOW

Once everyone has agreed upon the new flow, the technologists will build it. Once it is completed, it needs to be tested, and the best way to test it is to have the employees who have been performing the task run sample data and scenarios against the new work flow. When the system is stable and functioning, it can be released to the whole company.

STEP EIGHT: ANALYZING THE RESULTS

Once the new process is working smoothly, it's time to measure its success and define its impact. You chose to reconfigure this work flow because you identified a problem in it, so theoretically you have eliminated some amount of waste. It is important to measure what has been achieved because you can calculate the return on the technology dollars spent. In a big picture view, we often measure white-collar-lean efforts by studying such metrics as revenue per employee. The bottom line is that you should be seeing more efficiency and profitability as you strive to make the back office lean.

White collar lean is a powerful trend. You now have a good understanding of what it is and why it matters. Can you see how it might apply to your organization? I promise you it will continue to blossom over the years as competition continues to heat up and organizations struggle to do more with less in order to improve profitability and competitiveness. Those of you familiar with lean manufacturing know that organizations often bring in a "lean sensei" to help them optimize their processes. I provide the same assistance for organizations wanting to use technology to streamline white-collar work flows, so I guess you could say I am a white-collar-lean mentor!

Go to www.velocitymanifesto.com to contact us if you are interested in learning more about how we help organizations with this concept.

Now that you're familiar with the "leaning process," it's time to take your skills to the next level, that is, building powerful trendspotting and trend-leveraging processes as part of your high-beam strategy. We all have certain innate skills, but trendspotting is not one of them. It is a learned skill that takes an investment of time and energy. There is nothing magical in what I have done to develop an over-the-horizon vision that many consider valuable. You can do it as well, as you'll see in the next chapter. I promise you this; the return on your investment will be huge.

CHAPTER 13

BUILDING A HIGH-BEAM STRATEGY

Vision without action is a daydream.
Action without vision is a nightmare.

—JAPANESE PROVERB

WHEN YOU DRIVE DOWN A WINDING ROAD AT NIGHT, you rely on your headlights to see where you are going. If you didn't have headlights, you would run off the road, and the accident would be especially bad if you carried any significant speed. With low beams on, you can see ahead at least a little ways and can make progress without crashing. As you increase the velocity of your car, you need to see further ahead to anticipate any obstacles in the road since they'll be coming at you faster. How do you do that? By using your high beams. High beams illuminate the future by a few seconds more than the low beams do, and that makes all the difference. What if your high beams would let you see a mile down the road? You would be able to respond even more quickly and effectively to obstacles. You would feel a lot safer knowing you were not at the mercy of your reflexes (and your car's brakes and steering). The driver who uses high beams is able to drive faster and more safely than the driver who uses low beams. (And this type of driver is also more likely to leave high beams on when coming down the road facing me so I can be annoyed yet again . . .)

Business leaders run off the road every day because they haven't turned on their headlights at all. They can't spot upcoming trends and react to them quickly enough. The business leaders who adopt a high-beam strategy are the ones who are able to see furthest into the future, anticipate what's ahead, and push their organizations at high speeds. In today's high-velocity environment, low-beam leadership just won't cut it. Every organization needs high-beam leadership to beat the competition and thus survive.

High-beam leadership requires developing a high-beam strategy, one that builds on the ability to spot trends and react to them. This is not a complicated concept—the very definition of leadership is that you are out in front. If you have no idea what's coming next, how are you going to lead effectively? The leaders who can see the future most clearly, the ones who are using their high beams, are the ones who have the most success. High-beam leaders don't act in response to what they *hope* will happen; they react to what they *see* happening in the future.

I was chatting with the CEO of a medium-sized regional bank one day and asked him, "What do you think banking will look like over the next five years, and have you shared that vision with your team?" He shared a few ideas as to what might happen down the road, but he said he really didn't discuss them with his staff. So I asked him why he didn't put more energy into thinking about the future. He told me the subject just rarely came up. *Really? Seriously?* It just never comes up? That's exactly what it means to drive without headlights. Luckily, I was later able to help this individual and his executive team develop a high-beam strategy, and they quickly learned the importance of trendspotting and preparing for the future.

Lest you think I never run into forward-thinking people, please note that I constantly meet with entrepreneurs who have developed well-informed visions of the future and who are high-beam leaders. When I run into great visionaries, they often tend to be entrepreneurs. Every once in a while I do meet an intrapreneur—an individual who operates like an entrepreneur in a much larger organization—who has that wonderful eye for future trends. When I meet such people, I say a prayer that their leaders might see them for the valuable asset they are and give them the resources they need. These intrapreneurs can serve a valuable role as the high beams for their leaders and their organizations.

Intrapreneur: This derivative term is related to the word *entrepreneur,* with the differentiation being that the intrapreneur works within a large organization as opposed to independently. Intrapreneurs possess the same skills in that they have a desire to pioneer new ground, develop projects and products, or experiment with taking existing products in new directions. They just ply their trade within the walls of whatever corporation they work for at the moment.

As I have mentioned, high-beam leadership requires more than trendspotting. It requires being able to take advantage of a trend, or defend against its negative impact. I call this skill trend leveraging. While trendspotting is an analytical exercise, trend leveraging is an action item.

Think of trend leveraging as execution. World-class execution is a popular concept these days. Being able to execute against a plan is now seen as a very valuable skill, and I agree that it is. However, executing a plan that is not predicated on an accurate view of the future is deadly; executing a plan that is based on accurate trendspotting is the way to go. Yet much more energy is focused on driving execution for short-term gains than on successful trend leveraging. This mistake is punishing many large and well-known companies in the United States these days. An adjustment in priorities needs to be made because if you are on the wrong path, you can execute at an awesome speed right off a cliff—right, General Motors?

It intrigues me when leaders base all their decisions on current conditions and ignore the future. If times are good, they think the bounty will never end and rush to leverage their current operations until they are completely overextended. Then they get killed when things turn down and they are way too fat (as in the subprime mortgage crisis, for example). Conversely, when things are bad, they get depressed because they think the good times may never return, and often they just give up. When things look dark, they cannot see a future in which they can slim down, retool, innovate, and flourish.

A high-beam leader recognizes that economic conditions—the conditions of the road they're driving down—will change in the future, and that within certain industries there are cycles they must endure. High-beam leaders can see into the future and can keep their organizations from crashing. The execution-addicted leader cares only about the short-term numbers and monthly performance results and is blind to upcoming trends he should be preparing for. There were lots of travel agencies that were doing great—right up to the day Expedia hit the web.

Anyone can become a high-beam leader. The skills of trendspotting and trend leveraging are not supernatural. In fact, they can be reduced to a step-by-step process that anyone can institute with a little application of time and brainpower. I assembled and tested this process, which is explained on the following pages, by helping many organizations develop their trendspotting and trend-leveraging skills. Following these steps will enable you to spot trends and determine how they will affect your industry. In other words, this process will make you a high-beam leader.

THE HIGH-BEAM PROCESS

Before I run through the five steps of this process, let me share with you some observations from doing this process many times with clients. I have yet to ever do the high-beam process with a client when we were not surprised by two or three major direction changes that came out of the work. In every case, the executive teams I have worked with felt great at the end of the process if for no other reason than they were now all on the same page as to what they believed the future held. If you are doing traditional strategy processes today, please keep doing them. This is simply an add-on that may alter what you think is important today in light of what you decide will be critical tomorrow.

STEP ONE: Create a river of information, and use it

The first thing you must do is create a river of information. This river must consist of a broad cross section of input that can include ideas from knowledgeable people in your own industry as well as ideas from smart people in other fields. By ingesting their various opinions, you will be able to construct specific beliefs and see clear patterns. For example, if you are in retail clothing sales, keep a close eye on the clothing market, but also look at what is going on in retail electronic sales. Remember, there are three types of trends—organizational, industry, and macro—so you must create three respective rivers by connecting with sources that will provide information on possible trends of each type. For instance, you might subscribe to a group of blogs, Twitter feeds, magazines, and newsletters that contain articles about macro trends. Subscribing to industry magazines and newsletters will keep you up on industry trends. Internally, you must set up lines of communication with all types of employees in order to have your finger on the pulse of organizational trends.

This flow of information must be created carefully. It must be robust and varied, and you must be disciplined about tapping it daily. The key to creating a powerful river of information is getting the highest-quality content available in its most concise form so you have to invest only an hour or so a day studying it. The subsequent steps in the process of developing high-beam strategy will fail if you do not have a good river of information.

River of Information: The electronic flow of information that we have organized to flow into our brains. This includes using tools like RSS feeds, eNewsletters, blogs, microblogging, and social media sites as sources and transport vehicles to bring the information to us in real time.

STEP TWO: Trendspot with your executive staff

The next step is to set aside a block of time with the executive staff during which you go through the three types of trends and identify the top three in each area. Depending on the size of the organization, this could be done in a few hours or it might take a couple of days. I normally recommend at least a full day. This step is critical to the overall success of your high-beam strategy, so it takes serious thinking time—and the input of the team so you get a good cross section of opinions. This meeting should be held at least once a year—possibly more, depending on the speed of change in your industry. There should be plenty of healthy debate among your team—and the more people debating, the better.

Building a River of Knowledge

Please give this step its due consideration; if you are sloppy here, the rest of the process may be irrelevant. If you choose well with even four of the nine trends you identify, you may be ahead of where you are now.

STEP THREE: Tie each trend to your organizational strategy

Now your group needs to consider each trend individually and tie it back to the strategy of the organization in one of two ways: you will either plan to leverage the trend or defend against its impacts. You must then allocate specific resources toward the actions you decide to take. Your group should agree to one action item for each trend. Ideally, these nine action items should drive specific, measurable behaviors.

STEP FOUR: Take a broad look at the future

With this step, you want to take a look at the future from a different perspective. The group should now create a list of twelve ways in which the future—five years out—will be different from today. Identify major changes that may take place in people, the business climate, the regulatory environment, or buying patterns, for example. What you want to end up with is a concise list of the most important ways in which the future will be different. In some cases, items in the list might be extrapolations of the trends mentioned before. In other cases, they will be cultural or demographic changes. They could also be new uses of technology or alterations in corporate structures, including changes in leadership. Think of it as a science fiction project if that makes it easier! The point is to create a *shared* picture in the team's mind of what you will be dealing with in the future—whether it's positive or negative.

STEP FIVE: Develop an investment portfolio

Once you have created your list of trends and your list of ways the future will change things for your organization, you are ready to complete an investment portfolio for the future. Take the list of action items from step three and create a spreadsheet with three tabs to identify three different categories of potential investment for the future. These categories are People, Products (or Services), and Processes. Be sure you invest in all three of these areas and not put all your eggs in one basket. As a team, you now need to refer back to the trends you agreed upon in step three and develop specific investments you can make that will help the organization be relevant in the future you have been envisioning. By developing investments in People, Products, and Processes, you will have a balanced portfolio of investments you can monitor annually.

To access a full description of how we implement the high-beam process, visit www.velocitymanifesto.com.

Once you have a solid strategy for the future, it becomes *critical* to put energy and accountability into making sure that your organization is constantly making progress and that your team does not get distracted by focusing solely on near-term issues. Too many times I have seen companies build great visions of the future, write them down, and then fail to take action. As a leader, you must always keep planning for the future at the top of your mind. One thing is for sure: it's a lot easier to have an esoteric conversation as to where things will go than it is to make a public announcement to your team about how the organization will be preparing for the future.

In the heat of battle, simply having a great vision is worthless. Only when you actually take action to leverage that vision do you create value.

This is another one of the problems I have always had with some futurists: they make lots of money predicting what will come, but most of them do not start companies or make investments based on their own philosophies. I am proud to say I predict the future in front of audiences, and then I go and build companies based on my ideas and put my money on the line. There is nothing that refines the mind like having lots of your own dollars betting on your intellect.

Speaking of betting your own dollars, let's take a minute to talk more specifically about step three, in which leaders determine what to invest in each action item for the future. Every organization has a different ability to bet on the future, and every organization has a different level of risk or reward that can come from its specific situation. So, at one end of the spectrum, a company may be in a situation where it will be bankrupt in three years if it does not change dramatically because a developing trend is going against it completely. At the other end of the spectrum, a company may already be heading in a great direction and things are rolling along gloriously. In this case, the company just needs to invest in mitigating potential downsides and staying in front. Where you are along that spectrum will determine how much you need to invest, and where.

Next, you must decide on the specific investments you will make. Obviously, you are looking for investments that are high impact and cost-effective. In some cases, the investments will be expensive in order to have the possibility of a high return. Remember that you want to balance the investments across the three categories of People, Processes, and Products. This ensures that you do not invest too heavily in one area and set up a situation in which you develop a great product and have no ability to sell or service it. You want to also create a balance between long-term efforts and quick wins, as you would when investing in small-cap and large-cap stocks. They each have their strong and weak points, and the mix will balance risk and reward. Once you have completed and agreed upon the full set of investments for the next year, it is critical to communicate the plan to the rest of the team.

Be sure to use discretion when allotting resources in the People category; making announcements about rearranging positions in the future may cause needless or premature heartburn among employees. On the other hand, it is not good to keep the plan secret because you need the entire team to understand where you are going so they can make good decisions. You will also find that the team will be more excited and emotionally tied to the success of the organization if they know the plan for the future. It is the difference between just having a job and being on a mission.

I cannot overstress this point. I work with the leaders of some pretty large organizations who seem to believe their employees just show up to work in order to get a paycheck; because of this, management does not share *any* information about the company's goals or directions with the staff. This is insane to me. Keeping this information from employees actually de-motivates them. It causes your team to act like a herd of lemmings who just show up to do their jobs, without any collective passion for winning. What a waste this is, when we could instead put them on a mission they would have pride in and an emotional attachment to.

I am guessing that as you read this, you might be thinking, "This isn't me—other people make this mistake." Ask yourself this: Can you go to the receptionist or the warehouse workers, ask them what the company is trying to achieve, and get a clear answer? Try asking them what your market is going to be like in five years and see what they say. What's that? You don't think they need to know this information? Well, you might be interested in some of their observations. I am continually surprised at the perceptive feedback given to me by people who I sometimes think would be too young or too far down the org chart to really care about topics like market dynamics. Do not prejudge the value of a person's observations. There are many great stories about leaders who asked the lowest-ranking worker for an opinion and received the key piece of information needed. I was once told that Sam Walton used to ask the cart pushers and cashiers to name the dumbest thing the company asked them to do that day. Then he would take those opinions back to the home office and fix the underlying issue.

There are three possible outcomes to investing in the future:

1. The first possible outcome is that you are spot-on with your predictions, in which case you will of course reap the rewards of being well positioned in the market.

2. The second possible outcome is that you are completely wrong. In this case, you may lose the investment, but you will have learned what does not work so you can invest in something else next time. In this case, the best you can hope for is that you had a good learning experience.

3. The third possible outcome is that you are right but your timing is off. You bet on the change too early and are left with an investment that will not pay off for a

while. In some cases, you will lose this investment completely if you do not have the staying power to hang on until you can see a return on it.

You might have noticed that two of these three outcomes are not positive, and this will lead many people to claim that investing in the future is not worth the risk. I have run into many successful people who are convinced they have succeeded by simply being conservative and "sticking to their knitting." There are cycles in organizations, and in the later stages they have a choice: they can constantly renew themselves by updating their processes and strategies, or they can run out the curve using the same strategies that got them where they are. The problem with the latter strategy is that it eventually has an ending. At some point your competitors do figure out how to leverage changes in the market and then eat away at your market share. The market is just too competitive to keep your head down for years without looking up. In other words, there is as much danger these days in *not* innovating as there is in taking a few calculated risks.

There is middle ground that can be achieved; you don't have to lag behind your competitors or be light years ahead of everyone else. The answer lies in making affordable, strategic investments in this vision of the future you have created. Vision investing is much like any other financial investment you would make: the more you lock into one investment, the more risk you take. You need a balanced portfolio of investments in the present and investments in the future. You also have to accept that not all investments will perform at the same rate. In fact, some may be total losers. The trick is to win in the aggregate. And just as in financial investing, you must closely monitor your investments so you can move quickly when you see a positive or negative trend developing. If you make an investment in one of your ideas about the future and the trend starts to take off, put even more effort into it. If you are investing in something and getting nowhere or see that you might be wrong or ahead of the curve, slow down the investment or get out altogether. Knowing when to pull the plug on an effort is important in both of these investment scenarios, and so is knowing when to hold.

Once upon a time there was a young man who was steeped in the computer industry. He was an early adopter of email and was reading about the web back when "sites" consisted of just one page. He knew in his bones this thing called the Internet would explode into a vast and important tool, and he told anyone who would listen how big it was going to be. One day he was talking to one of his employees, and the conversation came around to domain names—specifically, it might make sense to register a bunch of words that someday might be hard to register. So they registered a handful that would be important to the business.

A few months passed before they had another talk about registering domain names—a vendor they knew had taken all of his money out of savings and was buying every single word he could find in his handy dictionary. He even wrote a program to check the domain listings and tell him what single words had not been registered yet. Our young man kept thinking that he should register names just because there were a ton of them lying around, but he never got around to it. He saw the trend, but he did not leverage it. His high beams were on, but he hadn't developed a complete high-beam strategy. That young man is older now and still beats himself up about this tragic error in judgment. What an idiot I am . . .

Don't make the same mistake I did. Develop a high-beam strategy that will allow you to become a high-beam leader. Keep up with your rivers of information, improve your trendspotting, make appropriate investments to leverage the trends, monitor those investments, and start the process all over again. By doing this, you will be able to drive your organization into the future at high velocity, and you will beat your low-beam competition.

PART THREE

CREATING A CULTURE OF VELOCITY

CHAPTER 14

THE TYRANNY OF THE ORG CHART

The only things that happen naturally in an organization are friction, confusion, and malperformance. Everything else is the result of leadership.

—Peter Drucker

HAVING GREAT DIGITAL PLUMBING AND A HIGH-BEAM strategy will not increase the velocity of your organization if you do not have the right people to get the job done. To be successful, you, as a leader, must learn how to create a high-velocity culture. The best tools in the hands of a disaffected team member will not produce results—period. I have witnessed organizations with great products and awesome systems in place fail because they simply do not have a culture of high velocity and strong performance. In the United States today, most people do not have to work for you—they have options and they darn sure will take another one if they don't feel valued, comfortable, or like they are making a difference in the world.

In order to play a high-speed game, you need team members who can work at a fast pace—and you also need a team that can work without friction. The high level of friction found in organizations today is often caused by a multigenerational workforce that has dramatically different views of the world and the tools used to operate in it. Individuals from different generations simply have very different styles of living and working. These differences can lead to a culture clash that results in dysfunctional teams. Technology has a big impact on this dysfunction, since team members from different generations prefer different tools and take completely different approaches to accomplishing the same task. There simply cannot be velocity where there is great friction. Your job as a leader is to create a culture that generates as little friction as possible by leveraging your employees' strengths and minimizing their differences.

Sounds simple, right? Just get everyone working together in harmony, highly motivated by a mission statement, all of them crystal clear on their role, and you have the velocity to accomplish amazing things. Unfortunately, this scenario doesn't happen much at all. We have four very different generations in the workforce right now, and another on the way. We have leaders from one generation who can't relate to the technologically sophisticated younger generations, and young people who have a somewhat warped view of what the organization owes them. We have completely blown up the concept of loyalty—both loyalty of workers to the organization and loyalty of the organization to the workers. We have leaders who have let us down strategically (General Motors), morally (Boeing), and financially (Enron). Bottom line: building a culture that is healthy—and hence can move with the maximum possible velocity—is very difficult at the moment.

In this final section of the book, I want to expose you to an array of ideas you can leverage to tune up your organization's culture. It will be more of a menu than a prescription; you can pick and choose the ideas that will work best for you. Don't think that having renovated your digital plumbing and developed a high-beam strategy is enough to lead your organization well, however. In the end, your people must be willing to use your digital plumbing effectively and move quickly according to your vision of the future, and these outcomes are accomplished by building a culture of velocity.

Put any group of people together for an extended amount of time and they will build a set of spoken or unspoken rules, mores, and expectations for how the overall group will operate. If you replace some of the original group with new members, you will be amazed at the influence the original culture continues to exert. In fact, even if you replace all of the people over time, vestiges of the original culture will remain.

I learned this lesson in spades when I took on the task of doing a turnaround for a company in 2001. After working on the project for three years, I walked away with one very important new piece of information: I learned I could not change a culture quickly by sheer force of will. I arrogantly assumed that I could instill a culture of velocity into a group of people who were hired by someone else and had years of inertia behind them. The good news is we got the turnaround done and the company is doing well to this day, but the lesson about the inertia of cultures remains.

As I hope you already know, the culture of an organization is absolutely critical to its overall success. Most organizations have been quite proactive in establishing a specific culture they know will benefit their mission. For example, Google and

Disney put a high premium on creativity and have instilled that into their cultures through reward systems. Other organizations have made customer service, strong moral values, or productivity their highest priority. Even the various branches of the military create specific cultures depending on their goals. The Marines have a different culture from the Air Force. This makes sense because the Marines are dropped in behind enemy lines at times and must be able to innovate solutions with no direction. The Air Force, on the other hand, is coordinating intricate missions that depend on people doing exactly as instructed. So, the difference in what it takes to get things done in each service drives their respective cultures. Consciously building a culture for your team is one of the most important factors in success. You can map out strategies all you want, but if you don't have a culture that drives success, you will never get there.

Younger generations will not operate like the current generation of leaders, and this is very important to understand. If you want to hire and retain A players, you will have to provide a vibrant culture for Generation Y and the rest of the generations to come. If you don't, you will find you will only assemble C players who can't get jobs in the world's top organizations. Leaders have the ability to groom cultures. As I learned, we can't just stamp our desires onto a culture quickly, but over time, we do have the ability to put specific practices and rewards systems in place and encourage mores that will turn the culture in the direction we choose.

Many leaders are running into problems as they build their cultures because they ignore the realities of the world today. They completely ignore how people are wired and the role of technology in the workplace. These cultures are huge battleships that are struggling to turn around as they head for icebergs. The cultural norms that made sense in the world twenty years ago just don't make sense now. How we communicate has changed; the hours we are available have changed; the ability to find information has grown; and we now frequently work in virtual teams. All of these dynamics have changed our cultural norms.

In addition, the pace of the economy continues to speed up, meaning there is an increased pace in innovation, product development, and information flow. If the culture of your organization is accustomed to a slower pace and most employees believe they have lots of time to complete each task, you will feel this cultural shift in your bottom line. For a culture to be healthy, vibrant, and successful today, leaders need to foster a culture that exhibits speed, creativity, teamwork, leverage of technology, and willingness to change. This applies to organizations as diverse as banks, churches, hospitals, governments, and schools. I don't say this because I think these cultural aspects are trendy; I say it because market and human dynamics are forcing us to take them into account if we want to be successful—whatever that means to your organization.

The great thing about digital plumbing is that once you build it and have it working, it takes very little upkeep. Not so for maintaining the culture of the employees who work with it daily. Technology can seem extremely complicated in many cases, but it is not nearly as complicated as getting a team of people to all head in a single direction and give 100 percent to get there. Until technology can run a company without human intervention (something science fiction writers predict, by the way), humans must interface with technology to get work done.

The things that motivated people in the 1900s and the 1950s are very different from what motivates them today. In the 1900s, people just wanted to survive. For them, working meant they would have food to eat and a roof over their heads. By the 1950s, people worked to get ahead. They worked to have more than their parents had, to have a nice house and a nice car. They could be motivated by the guarantee of a job and a consistent set of raises. Back then, the average worker would have three jobs throughout a career. It is a new ball game today. People are not just working to survive, and they are not just working to get ahead of their parents. They are working to feel fulfilled. They are working because they want to make a difference. The critical task for us as leaders is to find ways to get our people to give effort above and beyond the minimum amount they would have to put in to meet the basic requirements of their jobs.

In the United States, we are a predominantly white-collar workforce, made up largely of knowledge workers. We spend a decent percentage of our day interfacing with some type of technology. In order to create velocity in an organization, we have to rethink how people work with technology, and we have to develop better methods of getting four very different generations to adapt to the digital plumbing we are building.

We need to build cultures that are highly adaptive to the new technological tools being made available. Because technology is becoming such a high-impact variable in the success of an organization, it only makes sense that improving an organization's ability to incorporate these tools will have huge value. We all talk about the pace of change speeding up. We all talk about the benefits of adapting quickly to these changes. Yet, we seem to forget that in order to really get the full benefit of "change management" we need to integrate the skill of adapting to change into our organizational cultures.

Many organizations are learning that choosing, buying, and implementing technology is only half of getting it implemented fully and providing a healthy ROI. The latest and greatest set of world-class tools will be useless if a majority of the staff refuses to use them. As an outside observer, I am always fascinated by the cultural impacts of a decades-old company installing a fully integrated software system that takes three years to configure and impacts everyone. Jobs shift, power bases move, and work flows get completely changed. People leave, new people come on board, the older folks pine for the "old days," and the young guns get frustrated because everything takes so damn long to get done. New software acts as a catalyst to drive a substantial amount of change in an organization, namely, the retirement of the older generation and the promotion of the younger.

Because we have four generations all working together in today's workforce, there is a natural conflict in adapting to change. Typically, the younger you are, the easier it is to adapt to new tools and situations. Today, we also have the X factor of a huge inventory of new technologies coming into the marketplace every year. Technological tools are coming online so fast that the difference in comfort levels with these tools, even between generations next to each other, is extreme.

Just look at the changes in the devices we've used over the last twenty-five years. We saw PCs start to go mainstream around 1982, and for the next thirteen years they got smaller, faster, more portable, and more sophisticated. We also began to see a ramping up of software applications that provided many office tools that have now become commonplace—word processors, spreadsheets, presentation tools, etc. So the generation that was in its early twenties or younger in 1982 was lucky enough to build careers with these tools. The generation in its sixties had a hard road in adopting these tools. Its members had to slug it out in the market with youngsters who easily adapted to technology because they did not have years of experience learning to work any other way.

This generational dynamic is a whole book on its own, and it is not my intent to go down this road further. The point to be made is simply this: an organization's culture is made up of the behavior of the people inside that organization. The people who make up the corporate culture vary greatly in their comfort level with technology. If as a leader, you do nothing to deal with this dynamic, then whoever has the loudest voice will dictate what is acceptable in leveraging the new tools and processes technology brings.

I vividly remember working with a law firm in the early eighties as it tried to make the conversion from secretaries who typed all the legal documents by hand to an early version of the word processor. The young lawyers were excited about being able to use PCs to create their own documents and edit them at will, but the older lawyers refused to let go of their secretaries and typewriters. What is happening today in our cultures is nothing we have not already seen in the last three decades; the only difference is in the scale and pace of change.

Every generation is unique. Each has had a different set of influences and a different set of tools and technologies they grew up with. Every generation seems to worry about the one coming after it because they think they can see everything that is wrong with the younger generation coming up. We are now producing new technologies at such a rapid pace, the twenty or so years between generations is a lifetime in technological terms. Each generation is now growing up with a completely new collection of software and hardware to work with, be entertained by, and communicate through. It naturally follows that an organization would experience a lot of turbulence in deciding what tools to use and how to use them.

Your organization's velocity can be improved by leveraging the specific skills of each generation. The first battle is getting them to communicate and work together willingly. Special attention must be paid to the reality that the older generations typically control the budgets, and the younger need resources and the tools they are used to using in their personal lives in order to be innovative. This creates conflict when the younger workers feel stifled and ignored and the older generations feel skeptical and confused.

The new generation hitting the job market today has real gifts leaders must leverage. But the Millennials also have weaknesses. Experts have commented that the generation born between 1977 and 1998 is a coddled one whose members feel they deserve an easy road to reaching their goals. Their elders worry about their work ethic, because they often feel the Millennials do not have the sense of "paying their dues" that prior generations did. Millennials may believe the smartest person should be the boss, and in many cases, they believe they are the smartest because of their technological skill and ability to think outside the box. On the plus side, they can be wonderful entrepreneurs who are not afraid to be innovative and think on a global scale. They tend to be team oriented and regard their friends as family. They are culturally diverse and willing to accept people who are not like themselves as part of their team. Their sense of entitlement, although annoying to their elders, can drive them to success as they reach for what they believe they deserve.

Millennial: A member of the generation born between 1977 and 1998—often regarded by older generations in the workplace as entitled or spoiled.

An interesting fact is the rising number of job changes the Millennials will face. The U.S. Department of Labor speculates that by the time members of this generation reach the age of thirty-eight, they will have had ten to fourteen different jobs. Compare that to an average of three jobs a worker would have had in a career in the fifties. This is driven partly by the lack of loyalty businesses feel to their employees and partly by the fact that many Millennials do not work simply to make a living. They watched as earlier generations complained about their careers, and they have no intention of being bored or allowing themselves to be used in a way they feel is unfair. They have not seen a desperate economy, so they feel free to move between jobs for any one of many personal reasons. The bottom line is that many of them are talented, and they know it. They have bright ideas and many technological skills, and they want to use them. If they are not being fed emotionally and psychologically, they will shut down and want to move on. If you are a leader, you must understand and try to accommodate for the reality that Millennials have a different mindset from their baby boomer parents and grandparents, and this mind-set makes it difficult at times to get them into a situation in which they can flourish.

This is especially true with IT workers or employees who have a heavy dependence on computers in their jobs. A common example of how this plays out is a company that implements a new version of accounting software. There is typically a younger person in the accounting department who will grab the new software, help with the setup, and then eagerly learn how to use all of its tools. The older people in the department will lag for a long time before they learn all the features of the new software. If the organization doesn't reward the young accounting star for implementing and taking full advantage of the accounting software, she will simply move on to an organization where she feels more valued. I have seen this situation over and over in companies I've worked with.

When raising money from venture capitalists, one learns quickly that they strongly believe the number-one indicator of success is the strength of an organization's team. The team is more important than the business plan, and more important than the industry the business is in. They say, "We bet on the jockey, not the horse." They have learned that a good team with a bad business plan will recognize the situation and correct it—and still find a way to win. But a bad team with a great business plan fails almost every time. As we move toward a knowledge-based economy, people are becoming the most powerful factor in reaching goals. Sadly, many organizations still do not understand how crucial the quality of personnel is to success. They say they understand it but then sloppily assemble teams and fail to develop them. Worse yet, they may assemble a mixture of A and B players but then stick them in a culture that does not ignite their talents.

A great place to start rebuilding the culture is with the organizational structure. The old-school model of a layered hierarchy doesn't work as well as it did in the past. The younger people coming into the workplace do not like models that shield executives from newer hires. These younger workers have a gut-level understanding of technology and an absolute absence of fear when it comes to adopting new tools. This has created an environment in which there is conflict between the generations as they struggle to learn how to use the digital plumbing together. For this reason, many young people want to feel like they are a key member of the organization they work for; the last thing they want to see is a pyramid org chart where they are seven levels from the executive team.

So let's look at an example of a decades-old practice we can change to bring social benefits to a culture. We have three to four generations in the workplace, each of which is very different in its approach to using technology, and there is often a dynamic of young guns who feel it is their right to move up based on skills and talents, not tenure. At the same time, your industry is becoming more competitive and faster—and the pace of change is speeding up continuously—so you constantly have to reevaluate where you are going and how you are getting there. Add this all up and you will quickly see that a tiered, highly structured org chart is no longer the most effective framework for your company. Such a structure may work well for the military in some cases, or for very large, mature organizations. But even in the latter case, a hierarchical org chart doesn't work for smaller sub-teams. An elegant solution is to change the way we construct org charts, moving from a layered, top-down paradigm to a round structure that still tells people who they report to but uses a completely different model to do it. The graphic on the facing page is an example of what these "round" organizational charts look like.

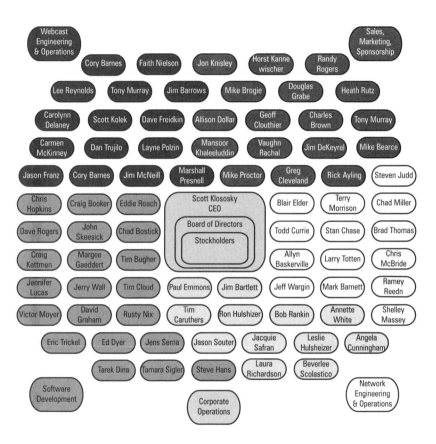

You might be thinking this concept is new and edgy, but Walt Disney actually used a round org chart to describe his company's structure in 1943. It isn't exactly the same paradigm I am suggesting, but it is awfully close. He followed the same paradigm of putting the management at the center of the circle, and the operating groups that interfaced with customers on the outside. He also delineated groups by creating pie slices in the circle. Walt Disney was one of the most creative businessmen ever to walk the face of the earth, and there is a lot to learn from him still.

The first thing you might notice is that these charts look more organic than the traditional pyramid, which is a good thing because people are organic and fit better in models that do not make them feel like building blocks. The round shape also delineates a wheel, an analogy that works beautifully throughout the chart. The people toward the outside of the wheel are the ones closest to the customer (where the rubber meets the road). They are also the most important people in the organization; without happy customers, you have no revenue. This may be hard for executives to accept at times, but if you truly understand the concept of servant leadership, you will love the thought that the hub of the wheel simply supports the outside, where the organization is focused on rolling. As you move inward, you delineate the management staff, until you get to the hub, which is made up of the CEO, the board, and the shareholders. The spokes, as it were, are the different divisions.

This type of org chart sends a very different message to the people who see it. Employees no longer get a sense they are many layers removed from the executives. They also see that the people closest to the outside are actually the most important, since they are the closest to the customer. Because this model does not emphasize top-to-bottom levels, people are more apt to carry ideas directly to the leaders of the company. This is critical for the Gen Y team members who want to be heard!

Shattering the tyranny of the org chart is not done by simply drawing things differently; it comes from actually behaving differently. In order to remove friction and to enable velocity, we need to constantly evaluate the teams we have and be ready to reorganize them to improve results. We need to be ready to mix up the generations so the strengths of all can be shared. We need the wisdom and experience of the older generations to be combined with the energy and technological knowledge of the young. We need to allow for the ad-hoc creation of smaller problem-solving teams that can then be turned loose to experiment. We need to allow employees to work in new ways with the IT department so they can find their own ways to improve their part of the digital plumbing.

Perhaps most difficult of all, we need to allow some of the sharp young people coming up—the ones who have great command of technological tools—to lead major efforts. Although they may not have all the savvy of their elders, they have a perspective too good to ignore. They also have short attention spans in most cases, which translates into an impatience that can be harnessed to drive velocity. Allowing younger staff members to lead in some areas serves a dual purpose: in addition to driving velocity, it also helps you hold on to your young A players. A perfect project for these young stars would be one focused on leveraging technology to achieve a goal that is in line with where the organization wants to move. Another option is using them as a stalking horse to experiment with new ways of accomplishing established processes. Their unique view on the world can often provide great benefits.

If you look at some of the hottest technology companies of the last twenty years, you see that many of them were started by a team of people in their mid-twenties. Think about the impact companies like Microsoft, Google, Yahoo, Facebook, and Twitter have had on the U.S. economy. Very young entrepreneurs started all of these. At the same time, in almost every case, there was a team of experienced executives who were brought in to grow these companies quickly. That combination of what each generation does well allowed these companies to prosper.

As older and younger generations work together in your organization, remember that they tend to view the very nature of technology in different ways. Most older executives I work with initially think the construction of technology is an engineering project. This leads them to believe that the people who build it are engineers in mind-set. In fact, we even use the term "computer engineering" to describe the assembly of networks or software. But the reality is that the writing of software and the construction of networks is an *artistic* process, which we'll discuss further in the next chapter. Not understanding the art of technology has huge—and not very positive—implications.

CHAPTER 15

TECHNOLOGY AS ART

An idea is salvation by imagination.
—Frank Lloyd Wright

THERE IS ONE SPECIFIC MOMENT IN MY CAREER AS A CEO that I will always remember. I walked into my office after yet another series of plane flights and plugged in my computer, intending to check my email. The only problem was that my Internet connection was down—yet again. The same thing had happened on several other occasions when I returned home from the road. Now understand, I ran a technology company with about one hundred employees (including network engineers), and I had told them that if I ever walked in and did not have a connection again, heads would roll. I stomped out into the middle of the work area and started yelling and spewing threats to everyone in the company. The digital plumbing was screwed up once again, and damn it, I was pissed off!

I'm embarrassed to tell this story because it may have been my worst moment as a leader. But it makes an important point: technology can be powerful and infuriating at the same time. I'm sure you've experienced similar feelings when your digital plumbing let you down. When you feel the frustration, the easiest thing to do is blame your IT staff. Unfortunately, this is exactly the wrong thing to do if you want to decrease the friction in your high-velocity culture.

Much of the friction in today's business cultures exists between the IT staff and leaders who do not really understand technology that well. One of the reasons it makes sense for you to learn more about technology in general is that it will help you work more productively with your IT staff and understand why they operate the way they do. The first thing you have to understand is that creating technology is more than just typing code. It is an art. The most talented people in your IT department are gifted visionaries who solve problems by relying on intuition and inspiration as well as technical knowledge. In order to get the most from your IT department, you have to learn to manage the staff as creative individuals, not just functionaries performing a well-defined task. Only if you understand the complexity and promise of your digital plumbing will you be able to work with your IT staff to optimize your technology. Not getting email may seem like a simple problem until you realize that reworking the email system involves restructuring a substantial portion of your digital plumbing.

The differences in viewpoint and knowledge between technologists and leaders has led to a cultural disconnect in many organizations. Over the last twenty years, I've seen many IT workers retreat to dark offices where they focus solely on keeping the complicated infrastructure up and running so no one yells at them. When they do come out, the business side of the house throws in requests for changes and new functionality without having any idea what that might entail. In the end, the IT people feel they can never really win—they have the very complicated job of keeping the plumbing running, but their leaders often have no idea what that takes. Almost without fail, this creates conflict between the IT people and business leaders. I see it again and again.

It seems like forever ago now, but it was just 1995 when I first met Marc Barker. I was putting together a team of programmers to build a marketing CD-ROM project my company had sold to a client in Washington, DC. Marc had been recommended as the perfect person to help us. I was to meet him at a local university, and as I walked into the room where he was working, I was surprised to see a fifty-year-old, gray-haired guy hunched over a computer with an unpleasant look on his face. A professor in the computer science department, who explained that Marc was overdue on a deadline for a multimedia project, introduced us. I had expected to meet a scruffy geek in his early twenties. Instead I got an artistic-looking elder who looked like he could have walked right out of a gallery. He proceeded to dismiss me with a wave of his hand because he was too busy to talk. After some convincing by the professor, Marc met with me and I contracted with him on the spot.

From that day on, we have been friends, and he has taught me quite a bit about the concept of technology as art. One of the most important things he got across to me about artists is that their skill starts with how they see things and ends with the technical capability to express what they see. He demonstrated this by putting a lamp next to a white wall and asking me what colors I saw. I answered, "White and gray," and was impressed with myself for throwing in gray—I figured most people would just say they saw a white wall. Marc proceeded to point out a rainbow of "cool" and "hot" colors he saw on the wall. Upon tilting my head like a dog, I finally saw what he saw. I have never forgotten this lesson, as well as the others he taught me later.

Somewhere along the line we started looking at technology in a dramatically incorrect way. Because the technology most of us touch is hardware (PCs, laptops, MP3 players, etc.) we fail to notice that the software that runs these devices has become as complex and beautiful as pieces of art. In fact, software is more complicated than is the most involved symphony, and has less tolerance for mistakes. Executives who didn't grow up immersed in programming, network engineering, or using complicated software applications seem to view the assembly of technology as a modern version of factory work. Programmers are often treated like the sweatshop workers of yesteryear (or, in some cases, current third-world nations). Understanding that the assembly of IT is artistic in nature forces leaders to understand they must employ great artists who have a talent for innovation in order to get results. We must also create a culture that nurtures the mind-set of an artist and acknowledges that, just as in art, there are many ways to accomplish a goal with technology. By viewing technology assembly from this more artistic viewpoint, you will be able to better build digital plumbing that will be productive, sophisticated, and unique.

In our modern, highly electronic world, we have come to believe that the inventory of technological tools available to us is all we need to get our organization where we want it to be. That viewpoint is dead wrong. Our inventory of technologies is more like a palette of art supplies—it *must* be formed into technological systems by true artists. Understand that ten different artists can be asked to paint a potato, and all ten will deliver unique paintings, some better than others. Similarly, ten different programmers can be asked to design an accounts receivable program, and one may do something that is awesome while many may code an application that barely functions. Why is this? Because artists all perceive instructions in different ways and have different skill levels. Traditional artists also have a large set of tools they use in putting a project together—brushes, paint, stone, canvases, etc. Programming artists are no different—they use many different languages, styles, and tools for coding. Each programmer may perceive what the organization needs in a different way.

By viewing technology construction as an artistic endeavor rather than a mechanical one, we are freed to not only build applications that have a stronger ROI, but also to do a better job of managing IT people. *Software engineers, developers, and programmers must be viewed as the artists of this generation.* The real thrill for IT professionals is the ability to be creative, to design and write something no one else has seen, and to earn the respect of their peers through innovation. I have never met a coder who works for a big corporation on boring, highly specified code who doesn't write artistic code on their own time just to keep their sanity.

Older managers often treat talented programmers who distinguish themselves by learning the business and then putting out valuable code as little but ditch-digging coders. Even though such programmers have good relationships with employees from other divisions and can listen to their needs and dependably translate the solutions into software, they are paid on a scale that looks like something designed by the military. Once managers see they have good laborers on staff, they give them even more work—often doubling the work assigned to the talented programmers compared to the mediocre programmer in the next cubicle. Then, when star coders ask for a raise (because they are smart enough to see the value they bring to the company), they are told that it will break the scale. So, talented programmers simply find a recruiter and move somewhere with better pay and a bigger cubicle.

On top of this, IT professionals are forced to rebuild and upgrade the digital plumbing while other people use it unceasingly. When something gets screwed up, the entire plumbing system comes down. Sadly, when things go right for months at a time, no one stands up and applauds. But once one thing breaks down and impacts a user, someone will be screaming for retribution.

Combine these dynamics of overwork and undervalue, and you see the gap between what leadership values and what leadership rewards. I am continually amazed at how far off base some executives are in understanding the difference in value between exceptional IT professionals and average IT professionals.

Many businesses make the fatal mistake of believing that IT is responsible for figuring out how to improve the digital plumbing *as it relates to specific business strategic goals*. In reality, most IT people are focused on keeping applications and data centers running smoothly and fully backed up. They burn countless hours answering help-desk calls that vary from the extremely complicated to the mundane. They aren't spending much time or energy brainstorming the next method of using software to reach departmental business goals.

Many times, this inability of the IT department to brainstorm new uses for existing software is the result of being understaffed or experiencing high turnover. IT departments are often understaffed because the leadership does not have a solid strategy for implementing technology. Without this strategy—and without ways to measure the ROI they might receive from it—they make decisions about the size of the IT staff based on how many people they perceive should be in it. This often results in an IT department that is overly scaled down or missing a critical member normally referred to as a software analyst. Without this person, there is no structured effort from an IT person to brainstorm new uses for the technology available to the department.

In order to keep up in the arms race of technology, you need to apply creativity and innovation as you assemble your digital plumbing and the team that will build and operate it. Software development cannot be done by the IT equivalent of ditch diggers. Maximum results are always found by mixing business-side experts, technologists who can construct, and geeks with imagination. Fail to add that last element, and you will have generic plumbing that might work, but isn't powerful. For that, you need to allow your geeks to act like artists.

Here's just one example of how this can be accomplished. I recently worked with a client who relied on a strict hierarchy of positions. One level of employee sits in a cube, the next level gets an office with no window, and management gets offices with windows. I noticed that this company's programmers were really struggling with concentration while working in open cubes where people were constantly walking by and distracting them. I suggested moving the young programmers to a room with a door would allow them to be much more productive. The executive I suggested this to didn't feel comfortable doing it because the programmers were "not management." He said that everyone in the company says they could be more productive in an office with a door. On the surface, his argument makes sense. What he didn't understand is how much productivity was truly being lost compared to other non-programming tasks. This executive has never programmed. He has never had to hold the complexity of thousands of lines of code in his head, only to have that shattered by someone dropping by to talk or making a comment while passing through. Failure to recognize small issues like this one causes huge productivity loss, which is a shame when it happens because of an easily changed HR rule.

When we give IT people clear direction on where we would like them to go, fully brief them on how the results will help the organization, and then turn them loose, we often will see the creation of fine art. If you are still not convinced, look at the IT cultures that have been built at such companies as Microsoft, Google, Oracle, Apple, Hewlett-Packard, and others. These companies dominate the market, and they do so by giving their technologists the freedom to dream and innovate.

Many manufacturing, financial, distribution, and retail operations have built IT departments that are forced to adhere to the overall company culture. For example, a manufacturer is typically very conscious of payroll costs because they need to pay the people who build the product as little as possible to maintain their margins. They need to have rigid processes to fit people into in order to maintain quality. They tend to be very careful about innovating, since problems with new products can be expensive. So the IT department for most manufacturers is assembled and trained with this mind-set. And predictably, the IT capabilities at manufacturing operations are not what they could be. This is obviously a very unenlightened method for setting a culture in an IT department. Regardless of how the rest of the business operates, the IT department must have a culture that incorporates at least some elements that allow for innovation, debate, and research into new ideas.

Understanding that technological improvement is an artistic process is extremely important for today's leaders. There will be an arms race for the next decade, and it won't be about how much equipment you can buy but how much IT talent you can assemble and retain. Technological infrastructure tools will become easy to buy and rent, to the point of commoditization. Software programs that handle accounts receivable, customer relationship management, and various types of reporting will become extremely basic tools most people will be familiar with and know how to implement. The hardware and networking components that create the skeleton of IT will be so easy to assemble, the need for network engineers will slowly decrease over the years. While much of what seems difficult today will become easier to implement, there will be a growing layer of coded intelligence that sits on top of the underlying plumbing. This layer of organization-specific software will be highly tailored to running each company and will decide the success or failure of many businesses.

Creating this layer of highly customized logic will be developers who are more software analysts than coders. These people will be knowledgeable about using technological tools, and they will know exactly how the organization runs. Their job descriptions will look very different from those of the developers we have today. They will be people who grew up with technology and understand it in their bones. They will know how to set up hardware and how to program, and they will be limited only by the time it takes them to learn about the organization's ways of doing business. Once they understand the operation, they will apply their skills to make huge amounts of progress, for themselves and the company.

We are starting to see this phenomenon in the young people coming out of college today. In order to win in the market, businesses need to invest in retaining the best and brightest, even if it means paying them large amounts of money that seem out of scale in traditional models. The trade-off for this investment will be the money they will drive to the bottom line with their skills. More and more, they will be like artists, whose canvases will be your business. Some of them will be able to do brilliant things, and they will be worth the money. Your job as a leader will be to spot these individuals, hire them, and manage them so they deliver to their fullest potential as employees and artists.

 If you're interested in more information on and examples of why technologists have an artistic mind-set, visit www.velocitymanifesto.com.

CHAPTER 16

THE POWER OF SOCIAL TECHNOLOGIES

Technology is shifting power away from the editors,
the publishers, the establishment, the media elite.
Now it's the people who are taking control.

—RUPERT MURDOCH

BEFORE I LAUNCH INTO A DISCUSSION OF SOCIAL technologies, I want to explain why I am including this content in this section about culture. One of the biggest differences in thinking between older leaders and younger employees involves the use of these web-based tools. "Social tech" is a powerful trend that is impacting how we communicate, who we connect with, how we distribute information and opinions, and—through crowdsourcing—how we get work done. It is also dramatically changing how we advertise, sell, and promote our products. It is creating a paradigm for online reputations that will make or break careers and organizations. It is not a fad—it is a powerful trend—and it is way past time for people to stop asking if they should be trying to learn about it.

Before I go any further, I want to clear up some vocabulary issues around social technologies, because many people
use different words to refer to the same things. I use the term
social technologies as the umbrella term for what has been called
"Web 2.0." Underneath that umbrella are three distinct areas.
The first is *social media*, which consists of services like YouTube,
SlideShare, Flickr, and Scribd. These are sites that let people
share media as part of a community. Then we have *social networking*, which includes MySpace, Facebook, Twitter, LinkedIn,
and any other service that is about facilitating community and
communication. The last area is *social relevance*, which denotes
anything having to do with the online reputations of individuals or organizations. In short, social relevance is what your cre-
dentials look like when someone searches your name online.

Social Technology: An umbrella term encompassing *social
media* (YouTube, SlideShare, Flickr, etc.), *social networking*
(MySpace, Facebook, Twitter, etc.), and *social relevance*—a
person or entity's online reputation.

In order to facilitate a healthy culture, leaders must learn
how to implement social tech institutionally in productive ways.
You cannot ignore it; it won't go away. You cannot control it;
it defies control. You must learn to use it to support teamwork
in all forms. Anything less is spitting into the wind. And I am
amazed at how many leaders are doing just that.

It is a shame that so many leaders fear various aspects of
social technology. They fear the invasion of privacy; they fear
employees wasting time online; they fear information leaks;

and they fear the unknown—all because they simply have not learned to use social technology in their own careers. A 2009 study showed that more than 54 percent of CIOs block their employees' use of social technologies (Robert Half Technologies, "Tweet This But Not That," April 13, 2009). In keeping with the saying that people fear what they do not understand, we have a generation of leaders who often have knee-jerk reactions to every new tech trend, which prevents them from seeing the beneficial uses possible from these new tools. Don't forget, not too long ago many executives demanded that IT departments block the entire Internet from employees, thinking it would just be a time-wasting playground. They have since fallen back to a position where they simply monitor usage in many cases. They were forced into this; the web became such a powerful and highly utilized tool, they simply could not restrict it any longer.

Let's get to the bottom line: Why does social tech matter?

1. So far it is the simplest and most efficient way known to mankind to communicate with others.

2. It gives users the ability to find others with similar interests, regardless of geography.

3. It gives everyone a voice. At no charge, any person can talk to billions of other people, and the readers vote with their mice as to whether the content is valuable or not.

4. Organizations are made up of people, and people crave the three things listed above.

Once people become proficient in these tools, they use them at work. You cannot stop this, and you should have no desire to do so. (Well, at least as long as they are using social tech for business purposes.)

I recently had a very large client that at first blocked Facebook but then reopened it to the company networks. When I asked why, this person said new employees (most of whom were of Gen Y) were so disgusted that the company would block a major technological tool, they would either not accept job offers or would leave quickly. This large, technology-oriented company looked like the Flintstones to new recruits.

I know what you might be thinking: *But Scott, what can you really do with social tech that is valuable to the organization?* Good question. I have a few answers. First of all, you can use these tools to build tighter relationships with your constituents, be they clients, customers, members, vendors, or otherwise. The great thing about social tools is they allow you to communicate with your contacts on a daily basis without much effort. I often use the example of the firm that helps me prepare my taxes. Once a year they mail me a packet and then send me an email exhorting me to send them my tax information so they can bill me lots of money for helping me negotiate tax laws that Einstein would pull his hair out over. I don't hear from them for the rest of the year. I would love for them to send me a constant stream of information on tax law changes and best practices for tax preparation so I could be efficient all year long. They could easily use social tech tools to do this for all their clients and in the end would have much tighter relationships with us.

Social tech can also be used to help you become recognized as an industry expert. In the past, you had to publish a book or become a speaker to do this, and even then many people would have no way of finding you. With blogs, Twitter feeds, newsfeeds, and other tools, you can now publish your ideas in a medium that is searchable; if you have important things to say, others will pass them on to their contacts, and your readership will increase. The larger your readership, the more influence you have when it comes to selling products or putting together partnerships. In short, social tech can help you be relevant in a market in which you might be invisible now.

Social tech can help you create a positive online reputation—personally and organizationally. Whether you understand that it can or not, you *will* have an online reputation. People will search for your name and your organization's name to see what others are saying about you, and they won't be nearly as interested in what you are saying about yourself. Why? Because, according to a variety of studies, roughly only 13 percent of people believe advertising, but nearly 80 percent believe what other people tell them. From this day forward, you should be doing everything you can to construct a positive online reputation for yourself and your organization.

Fortunately, there is an easy three-step formula for enhancing an online reputation. Step one is to build a listening program. This means running alerts and monitoring applications that will notify you any time your name or the organization's name is used on the Internet—especially in the social sphere. Step two is to develop an engagement policy. This is a written set of instructions on what actions need to be taken when a positive or negative comment is made about you online. Step three is a measurement system so you can gauge how often you are mentioned, and what the ratio between positive and negative comments is. Every organization, and anyone in management positions, should be running an online reputation management system—even if it's just a basic version.

In the socially enabled world, it is not wise to try the "ostrich with its head in the sand" approach. If you'd like a list of tools to help you monitor your online reputation, and to follow what people are saying about you and your organization, visit www.velocitymanifesto.com.

Social tech can be used to find unsatisfied customers so that you can attempt to remediate the relationship. Many organizations, large and small, are monitoring the Internet for people who mention their name negatively. Once an unhappy customer is discovered, they have organized response systems for trying to get them back on the happy side. As online reputation becomes even more critical in people's buying decisions, you need to pay more attention to the online chatter to address negative comments. Social tech is great for consumers: for the first time, we have a way to tell the world when we are wronged. We can share our feedback with millions of people instantly and at no cost. Businesses might think this unfair at times, but in the end, it will make for much better customer service. Companies will no longer be able to abuse a large customer group and think they can get away with it.

Social tech can help you lower back-office costs by utilizing crowdsourcing to get work done by the Internet herd. The use of crowdsourcing is exploding, and companies can do everything from having logos designed to outsourcing R&D to gathering free market research. If you think outsourcing has had an impact on businesses, just wait until crowdsourcing as a normal model for getting work done explodes. You need only look at Amazon's Mechanical Turk (mturk.com), Dell's Idea-Storm (dellideastorm.com), or Innocentive (innocentive.com) to see where we are headed. For many organizations, massive benefits could be had today if only they understood how simple and powerful crowdsourcing is.

As a leader, you have to lead your organization's culture toward the use of social tech. This will require getting some of the older members of your organization to adopt something new while getting younger people to focus their use of social media on activities that will benefit your organization.

When thinking about what social technologies can do for you and your organization, it is critical to apply the knowledge and perspectives from some of the earlier chapters in this book. Understanding your digital plumbing will help you see how disseminating information to your constituents through social tech tools could create much tighter relationships between them and your organization. Learning how to apply the DIKW chain to the conversations and information provided through social tech could give you a finger on the pulse of your customer base that you have never had before. For example, aggregating positive and negative comments and then analyzing the possible dynamics behind why you are getting each type of response from a selected group could yield powerful information about what you are doing right and what you are doing wrong.

Social tech can be used to build a river of information into each employee's brain. This river of information is essential for all organizations, as discussed in the section on trendspotting. We must all sign up for newsletters and blogs that relate to our fields of endeavor. Today we can follow experts from any industry on Twitter and set alerts so we see new Internet content on topics that interest us, and then filter all of this information onto one screen. If leaders would set an example by doing this themselves, and mandating that all employees do some variation of it that applies to their position, the organization as a whole would be a lot smarter. And being smarter usually translates into more profits. You can build rivers of information on

your industry and your competitors from thought leaders, content providers, the government, and customers. I spend about
forty-five minutes a day reviewing the stream of information
I have created. I cannot even imagine how much less I would
know had I not done this over the last few years.

Using social technology to advance your organization is
such an important part of success that I have given it a name:
enterprise social technology. Enterprise social technology must be
part of your culture. There is obviously a huge benefit in learning how to apply a powerful technology trend like this earlier than your competitors do. It means, of course, you have to
build some velocity. Velocity comes from having a plan—you
don't have time to experiment with best practices for a year
or two.

Enterprise Social Technology: This term refers to the technology tools and concepts that can be leveraged by organizations as opposed to the personal use of social tech tools.

The reality is that today, most organizations are in a state of
chaos regarding how employees use social technology. Because
these tools are so new, you can often be part of organizing
their first use in your company. Rarely in a career do you get a
chance to implement a brand-new strategy that is dependent
on creativity and discipline, and in which experimentation is
the norm. The faster you can help your team leverage these
tools, the more of an advantage you will have in the marketplace. Move too slowly, and you will learn the hard way why
high-beam leadership is critical in this day and age. Neglecting
or avoiding social tech will also disenfranchise all the young

people who have come to love and depend on it to operate their lives and their careers. Don't put off finding ways to use social tech in your organization—it's a set of tools simply too beneficial to ignore.

I have developed a holistic plan for helping organizations move from very little use of social media to integrating social media deeply into their overall strategy in effective ways. To access this step-by-step guide to integrating social technology into your company's overall strategy, visit www.velocitymanifesto.com.

CHAPTER 17

SOCIAL TECHNOLOGIES AND THE CHANGING PRACTICE OF SALES

*The death of the salesman has played on Broadway
for years. Now it is becoming a reality.*

—Scott Klososky

(OK, I HAD TO USE AT LEAST ONE QUOTE OF MY OWN.)
No process lies closer to the heart of an organization's culture
than the selling process, and there is nothing that causes more
friction between generations than determining how to close
sales. One of the most dramatic results of the social tech movement may be how it will alter the way we sell everything from
consumer goods to sophisticated, high-end services. Its impacts
will be felt at advertising and marketing firms and will be even
more important in the sales departments. The issue now is that
the old guard has been selling in a certain way for decades,
while the new guard has a whole new bag of tricks. Leading an

organization toward a selling process that incorporates rapidly evolving technologies will help you reduce friction between the "two martini" and the "online gaming" generations.

I was fortunate enough as a young businessman to have gone through sales schools at Xerox, IBM, and 3M. These were some of the finest programs in the marketplace at that time, and I am thankful I got to experience formalized sales training in my early twenties. I then gravitated toward the sales process for the rest of my career. I still get a charge out of identifying a prospect, building trust, nudging the process along, and closing the sale. I have followed a number of the trendy concepts in sales over the last twenty years just so I didn't miss out on some new technique that could be critical to driving revenue. In my career, I have watched the change from relationship selling to solution-oriented selling to process-driven selling. We now stand on the brink of a totally new sales process that is very different from anything we have seen before, and it is driven by the social technologies we have been discussing.

Every organization sells something. Social tech is beginning to have a serious impact on what that means. From an organization's online reputation, which a prospect can now discern on their own in about five seconds, to the methods by which the seller builds trust with a prospect, the entire sales process is being turned on its head. This has a huge impact on the culture of an organization because "selling" has always been one of the most critical functions in the operation, hence the old saying, "Nothing happens until someone sells something." The entire relationship between the organization and the consumer is shifting. The consumer wants to be able to communicate with employees of the organization at will, and the consumer will punish the organization publically if it does not live up to expectations.

Here is a wake-up call: the culture of your organization has never been more public than it is in a world where social networking allows anyone to talk to anyone else, and where everyone can see how people feel about your organization with a simple search of social networking discussions. The process of selling is now, more than ever, a team process during which your entire team will be on display. No longer will a buyer's contact with your organization be limited to a single salesperson and a barrage of advertising.

If you think of products along a continuum with consumer-oriented items at one end (toothpaste, staplers, candy, etc.) and sophisticated, complicated items at the other (consulting services, large software applications, manufacturing tools), we can start to frame how social tech will impact revenue generation. On the consumer end, we will need to employ "socially directed buying" in order to create the "eWord of mouth"—the electronic version of your customers telling other people about your product or service—and online promotion that will drive consumers to purchase at the local store. At the other end, we will need to employ "socially facilitated selling" in order to help a sales force close these big deals.

Socially Directed Buying: The use of social technology to build a strong relationship between your organization and the buying public. This term relates directly to the use of web-based social tools to help create business-to-consumer (B2C) revenue growth.

Socially Facilitated Selling: The use of social technology to help a sales force close complicated sales through industry awareness, online relationship building, and online reputation management. This term relates directly to the use of web-based social tools to help create business-to-business (B2B) revenue growth.

These two terms encapsulate a collection of social tech concepts that are influencing the success or failure of driving revenue. Socially directed buying is all about using social tools to do branding and marketing of products to the masses so that they choose to buy them when they walk into a retailer. It also includes revenue-on-demand schemas that help everyone from airlines to restaurants sell excess capacity to the public with instant discounts delivered over real-time social tools. Socially facilitated selling is a strategy of using social tools to help a sales force be better able to close complicated sales. This includes teaching salespeople how to manage their online reputations, build rivers of information about their industries, and tighten their online relationships with buyers. Defining these terms in more detail is the job of another book, but suffice it to say we will need to rethink "sales" in the upcoming era.

Customers can now find out much more information about our products and services than what is posted on our websites. They can easily search for what other people have said about us and find our customers and chat with them online rather than speaking with references we gave them (the ones who say only nice things). Depending on the products you sell, this change in customer behavior will impact you differently. One thing is for certain: salespeople cannot pretend to be industry experts when they aren't, and websites cannot get away with spinning the truth in a world where the authenticity of both these situations can be verified with a few mouse clicks.

Your organization must treat sales as a social interaction rather than a hunt. Social technologies now provide a way to have an ongoing conversation with prospects and customers. Without having customers visit the website, the organization can "talk" to its prospects and build up trust by providing valuable information—daily, in some cases. This more frequent communication gives the seller an open door through which to provide valuable content, coupons, or advice. The ongoing conversation also helps sellers stay at the top of their customers' minds.

In addition, social tools can be used to identify prospects by tapping into the detailed demographics now available through some online communities. Rather than depending on the shotgun approach of traditional advertising, social technologies allow for tight demographic and psychographic targeting. The cost of this laser-like approach is a fraction of the cost of traditional advertising—and the results are often much better.

Social networking creates fertile ground for eWord of mouth. Since we know that 80 percent of people generally believe what they are told by a friend, eWord of mouth is the single most effective driver of customer loyalty. Finding ways to facilitate happy customers talking about your products not only creates new customers; it also floods the web with positive chatter that will build your online reputation when prospects do searches to check out your organization.

Tools like LinkedIn and Plaxo can benefit salespeople individually by allowing them to stay in contact with a much larger number of people than was previously possible. When telephone calls and in-person visits were the only ways to stay in touch, it was only possible to keep up with a limited number of prospects and customers. But with the aid of social networking tools, it is now possible to keep tabs on thousands of people at a time. It is also possible to communicate with the same number of people at the press of a button.

All of this combines to create a whole new customer relationship model that is driven in part by technology channels between buyer and seller. The longer you hold on to the traditional sales model without integrating electronic tools, the more danger you face in having a drop-off in revenue. The best thing an organization can do is lay out a support strategy that helps the sales staff by integrating the positive aspects of this new customer relationship model while minimizing its possible negative impacts.

So, what would that support strategy look like? In a socially facilitated selling model, you would seek to provide as many web-based connection points as possible for delivering streams of information about your products and services. This could include blogs, Twitter feeds, newsletters, etc. The organization would take some responsibility for nurturing the relationship by investing in valuable content that earns the right to have a constant conversation with a customer. Keep in mind, customers and prospects do not have time to read through a river of marketing hype. They want content that is valuable to them, because it is fresh, new, insightful, funny, or clever. You can deliver marketing information through social networks as long as it is embedded in the stream and not dominating it.

You must also acknowledge that the role of salespeople is changing and that they might have to play various parts based on how the prospect is used to getting information and purchasing products. Some prospects will need very little hand-holding, while others may want to work through all the traditional steps from the past. In any case, you will do yourself a disservice if you try to take away commissions on sales that are partially supported by the organizational stream of information. Unless you really believe you will not need a given salesperson to play a role going forward, don't start taking sales and commissions away just because some of the traditional sales steps are moving online.

As mentioned in chapter 11, it is critical that organizations—and salespeople—learn to develop robust rivers of information. In order to continue to provide value, they must be aware of everything a competitor might be doing. They need to know every detail about the happenings in their marketplace, and they need to know details about their customers' organizations. All of this information can be gained by using a system of alerts and subscribing to various RSS feeds and newsletters. In a world where prospects and customers can access almost unlimited information at the touch of a button, a salesperson looks foolish, and in some ways worthless, when their customers know more than they do.

In addition, it will be routine for a prospect to search the name of her salesperson to see how relevant he is in his field. Many people are getting jaded about being "sold" and would prefer to deal with someone who is an expert in something other than the magical seven steps to a closed sale. We live in a new world where single people go home and Google the name of someone they just met, and what they find will have everything to do with the likelihood of a first date. This same generation will think nothing of searching for information on a salesperson to see if he has a good reputation or even knows anything about the products he sells.

Selling is all about the combination of each of the following three dynamics. One, you must have a product someone wants. Two, that someone has to have the resources to purchase it. And three, that someone must trust you enough to do business with you. One of the major factors in building trust from now on will be how prospects look at what Internet users have said about you. Even one negative comment can be devastating. I often have people ask me what they can do to defend themselves against unfairly negative comments, and there is no easy answer to that question. There is no way to erase what other people say. There is no court that will force people to erase what they have written (unless it is libelous). The best thing to do is reach out to the critics privately and try to get them to change their mind so they will willingly take down the information or post something more positive. The other tactic is to make sure that the positive words written about you outweigh the few negative reports. Then, at least on a volume scale, you will look OK.

Almost every organization depends on its clients, customers, or members for survival. What could be more important for a leader than the relationships the organization has with these constituents? Without a working knowledge of the advantages and disadvantages of social technologies, leaders are flying blind in one of their most important areas of responsibility. The solution is simple. Invest a little time learning to use these tools with your own hands. Be aware of how they are being used within the organization, and know how your competitors are using them as well. This is not a case of what you don't know not hurting you. You must support a culture that embraces these new tools and paradigms, and if you do, your company's reputation and ability to sell will blossom. Restrict usage of these new tools, and you will lose your brightest team members and cripple your connection with your customers. Don't be a leader who fears the unknown and therefore blocks the usage of a tool that has significant potential for your company. If you don't understand the tools, invest the time to actually learn how to use them, or get some reverse mentoring from someone who does. The rewards are great, and the investments small by comparison.

For more detailed information on how social technology will affect the sales process, visit www.velocitymanifesto.com.

CHAPTER 18

THE POWER OF VIRTUAL TEAMS

The prediction I can make with the highest confidence
is that the most amazing discoveries will be ones
we are not today wise enough to foresee.

—CARL SAGAN

WHEN WE FIRST DISCOVERED THAT PERSONAL computers could assist someone in working from home, I suspect we did not foresee the ultimate outcome—the move toward virtual teams. Slowly and steadily, we have broken down the model of all workers residing in a single location and working side by side. The use of home-based workers is being combined with outsourcing—and now crowdsourcing—to result in a new model for getting work done.

Yet again, technology has spawned a major change in the business world by allowing team members to collaborate across great distances. Wonderful efficiencies come with this capability, and it certainly allows for more velocity in assembling teams and then reassembling them when new conditions or projects arise. The ability to lead virtual teams is yet another example of a new skill that a modern leader must have. This type of leadership has unique dynamics that do not exist when everyone is side by side in the office. Do it well, and you create lots of speed and efficiency in leveraging talent. Do it poorly, and you create horrible inefficiencies that will drag you down and force you back into the traditional model of working. Strangely, there has been little attention paid to the best practices for leading and assembling virtual teams.

From a cultural standpoint, virtual teams create an interesting dilemma, since culture is normally formed by people coming to a tacit (or not so tacit) agreement on mores by interacting with one another. When people do not physically work in the same place, how do they develop their views on what is acceptable in their team's culture? The only solution is a new, modified form of leadership attuned to the dynamics of virtual teams.

I remember, back in the eighties, the first time I had the option to let someone work from home. I had an accounting person who was getting ready to have a baby. I didn't want to lose her and she didn't want to lose her job. I owned a handful of computer stores, so we had the ability to easily send her home with a computer and let her connect remotely into our office systems. She and I discussed how this would all work, and without a major shift in strategy, we easily allowed her to work from home for a few months. Even with a newborn taking up most of her time, she was able to keep the accounting reports and taxes rolling, and the setup worked out for both of us.

Not too long after that, we started to use more than the telephone to communicate between the computer stores. We started sharing data and uploading transactions and accounting information to a central computer. We started giving each store visibility into the service calls of the other stores. Without much planning, we became more of a team rather than five separate stores in separate cities. I went on to start other businesses and ended up with offices all over the country; we used videoconferencing, portals, instant messaging, and a slew of other technologies to make sure that one hundred people spread out all over the country still operated as a tight team, even though they rarely saw each other in person. I've been incorporating virtual teams into my companies for nearly twenty years now.

I was then lucky enough to get hired by Lockheed Martin to provide training on best practices for virtual teams. This gave me an opportunity to really study the field and find out what some of the large institutions were doing. I was surprised to see the extent to which organizations were already leveraging virtual teams. I was also able to interview a number of people and get a big list of all the things not to do.

If you want to build an organizational culture that moves with velocity, learning to facilitate virtual teams is a must. A part of this is just learning to use technology in the best ways to support these teams. A larger part of the successful management of virtual teams has everything to do with how you assemble them, and how you lead them.

Virtual teams are simply a different animal from the traditional team that works in one office. The dynamics are vastly dissimilar, and yet many leaders have given this fact very little thought. They allow virtual teams to be assembled as if they are creating a recipe for success using people as ingredients. They find the best finance person, put her on a team with a great marketing person and a wonderful product development person, and think they have a winning group. Then they learn people are not ingredients and the distance between them can be a huge barrier that creates much drama.

After years of struggling to build virtual teams that perform well, I finally learned there is a real science to selecting, managing, assigning work to, and rewarding virtual team members. I will cover in detail each of these areas in a minute, but before we go there, let's talk about all the reasons virtual teams are different from traditional, office-based teams. For simplicity's sake, I am going to shorten "virtual team" to "vTeam." You might think this is because I am too lazy to write "virtual" over and over—OK, you're right.

PEOPLE DEAL WITH BEING ALONE IN DIFFERENT WAYS

Depending on their personalities, people will deal with working remotely in many different ways. Introverts and extroverts both have strong and weak points in the virtual world. Introverts get energy from being alone, so we can work on our own for months and feel lovely. I say "we" because I am a raging introvert, and I can work from my home office for days without seeing another person and feel like I just had the best time in months. My wife, on the other hand, is a serious extrovert and does not like being home alone very much at all. Introverts can disappear on a virtual team because they often find their hole, crawl in, and work without keeping the rest of the team up to date. Extroverts will do their very best to build community with the rest of the virtual team because they like interaction.

These dynamics still hold true even when a person works at a different facility from the rest of their team. Even though people who are not on their direct team may surround them, they are not relating with the people in their specific group, so their personality profile may cause a problem. Now, I know some of you are thinking that personality profiles can cause issues even on teams housed together in the same office, and that is certainly true. However, I have found that any issues you would have in person are exacerbated on virtual teams when you add in the catalyst of distance, which is never helpful.

COMMUNICATION CAN BE MISCONSTRUED OVER EMAIL, TEXT MESSAGING, INSTANT MESSAGING, TWITTER, ETC.

Anyone who has ever had an email problem with a significant other knows what I am talking about. Words on a screen can be interpreted many different ways, and learning to use electronic communications is an art many people don't understand. We have people who email five-page missives on a trivial subject and people who send five-word emails to answer an involved question. I don't even want to get into how many times my wife and I have emailed something that didn't land the way it was intended, so I'll just tell you about a guy I work with now who has yet to type an email longer than twenty words in his life. He often responds to emails I send with one-line comments like "I don't agree," or "Let's fire him," or "You do it." I get frustrated when I receive these emails, and I usually have to call him to find out what he really means.

When a team of people works together remotely, its members naturally end up doing much of their communication over the wire. Sooner or later, someone will say something in an email or instant message that will come across negatively, even if it wasn't intended that way. Because the rest of the team members cannot see the look on a face or hear a tone of a voice, they may infer an emotion behind the communication that wasn't really there. The situation then devolves into a lot of wasted energy as people kibitz about why the author of the email or text message has a problem. Rarely have I found a virtual team loaded with great over-the-wire communicators. There are always a few people who cause a lot of drama, and the sad thing is that many of them really do not mean to offend.

I often run into a specific type of communicator who is a particularly serious problem on a virtual team: the "email flame artist." This is the person who would never say a heated word in person but uses the remoteness of electronic communication to justify venting strong negative emotions over the wire. They are easy to spot because it is clear they make an art form out of spending five hundred words eloquently explaining why their target has the dumbest idea known to man. This type of behavior is toxic to a virtual team.

IT IS HARD TO INSPIRE PEOPLE FROM GREAT DISTANCES

One of the important jobs of a leader is to inspire employees to reach goals and grow as team members. Inspiring people is an exceedingly personal task, and to do it well a leader must use different methods for each person. I have also found that inspiring people is easier to do in person than remotely. Imparting inspiration is especially hard when you want to bring a team together and give a rousing speech to get everyone pumped up to go out and slay the dragon. As the leader of a virtual team, the best tool you have is a conference call or videoconference. (God forbid you ever have to give an inspiring speech over email.)

IT IS HARD TO BE CREATIVE OVER THE WIRE

On just about any team, there comes a day when you need to do some creative thinking—or what I call "green light thinking." Nothing is more awkward than trying to be creative with a group of people over the phone. Most people are visual and want to see ideas on the wall. Technologies that allow teams to do group whiteboarding on a computer are available, but even this can be much less creatively stimulating than a physical whiteboard surrounded by a group of people. Creative sessions work best when people can integrate and expand on others' ideas, and when their minds can be put in a creative place. It is extremely difficult to get either of those done when a team is not together in a room.

IT IS DIFFICULT FOR TEAM MEMBERS TO "SEE" THE PROGRESS OTHERS ARE MAKING

When a team works together in an office, it is easy to stay current on how all the members are progressing with their respective tasks. Either the progress is visible, or the person gives updates around the office or in the lunchroom. When they're all in the same place, people can quickly get together and compare notes on how far along they are and get coworkers to review drafts of their work. Leaders can easily keep track of certain projects and team members because they can hover, check in, and get quick updates. The same cannot be said for virtual teams. When people are not working in the same physical space, they do not have those moments to quickly update each other. In many cases, team members have no idea how much work has been done on a project until the day it is to be turned in. This causes lots of problems because vTeam members begin to lose confidence in other people who work at a slower pace or who routinely miss deadlines.

Leaders have an even tougher battle as they keep up with their employees' work. They have to fight to get information on the status of projects, and often they do not find out about problems until it's too late. I frequently struggled with people who assured me they would hit a deadline and then would show up with half the work done on the day it was due. Had they been sitting in the same office with me, I would have known I had a problem.

MEMBERS OF A vTEAM HAVE A HARDER TIME BUILDING TRUST IN EACH OTHER

Human beings are funny about extending trust to each other. We have learned over the years to distrust people we do not know. Someone who talks to us on a conference call once a week and emails us every once in a while is not automatically our friend. Yet trust is one of the key ingredients in creating teamwork. If members of a team cannot trust each other to express how they feel or share their ideas, they will not be very productive. They will burn lots of energy on issues that simply waste time. They will worry more about getting credit, internal politics, and the motivations of the people they are supposed to be working with.

IT IS HARDER TO INSTILL A CULTURE WHEN PEOPLE ARE REMOTE

The reality of building culture into an organization is that it spreads to each member of a team virally. It is not something we can lay out in a handbook and expect people to adopt. Take it from me—I have been naive enough in the past to think it could be done this way. A culture is adopted because a group of people is led to—and then chooses to—behave within a certain set of parameters. When new people are added to the team, they either adopt the parameters completely or pull the culture slightly toward their own behaviors. On a vTeam, there is not enough interaction to easily transfer culture from leader to members, and from members to each other. It takes much longer to get a specific culture in place—if it can be instilled at all. If the vTeam is made up of people who have worked for the organization for a long time but live in different places, instilling company culture is not too hard. But more and more vTeams are made up of people who have never worked together, and they may include a few contractors mixed in with full-time employees. In this case, colliding beliefs about what is acceptable can be a big problem.

Here is a simple example from my career. I had a virtual team of people who worked across many different locations in the United States. Our home office was in Oklahoma, where I lived. The president of the company worked out of a small office in Los Angeles. He had worked for us for a few months but had never come to the Oklahoma office. The first time he came to visit the troops, he called a meeting to introduce himself in person. He had communicated with many of these folks over the phone and through email for months, but in person he made a huge cultural mistake: he stood up in this meeting and dropped numerous cusswords in front of the whole team. The Oklahoma employees had a culture in which it was not acceptable for someone to be swearing in a group setting, and it caused quite a stir. I had to take our president aside and explain to him we didn't do that in our culture. To his credit, he uttered almost no cusswords over the rest of his stay with us.

This list of their differences from office-based teams doesn't mean I don't love the concept of virtual teams—in fact, I was using them years before they were fashionable. But I have learned some hard lessons along the way. To this day, I look for the best people I can find—wherever they are—and get them on the team, even if they cannot move to a central office. I see a future in which it is very normal to assemble teams from the best experts you can find anywhere in the world and then dissemble the team as soon as a project is done, only to form another team for the next project. I also think we are going to see many more people eschew the typical on-site job route and choose to be independent contractors.

We are seeing this trend in the technology industry in spades. Many programmers and engineers are choosing to contract with a client (or clients) instead of jumping on the payroll. I am sure we will see this trend continue with accountants, lawyers, doctors, and other professions in which people align themselves with an organization for only as long as it makes sense, and only to perform certain services. The better you are at what you do, the more likely you are to choose not to be on the payroll. You may be able to make more money and have a more challenging career by being a freelancer. The laws of supply and demand will ensure that you always have a place to apply your talents; there will always be a need for top talent.

At the same time, organizations are finding that vTeams are a great way to lower infrastructure costs and avoid having to supply an office and infrastructure for their teams. They can also easily scale up or down a partially contracted team, and they have the ability to reach out for specialized talents and add them to the team without having to employ that talent permanently. There are many reasons why it makes sense to have your teams become more and more virtual, and we now have wonderful technological tools to help mitigate the negative aspects I listed. But as mentioned previously, technology can only improve parts of these problems. To leverage vTeams, leaders will have to develop new skills and approaches. In order to help you in this endeavor, I have listed on the following pages some hints and insights into how to build a vTeam that can achieve goals with velocity. I hope you will be able to put them into practice with your own remote employees or contractors.

ASSEMBLING THE vTEAM

The number-one consideration when assembling a vTeam should always be each member's skill set. We need people to do specific tasks, so if you need a marketing person or salesperson, then that is the first criteria. Following this closely is the member's personality. As I touched on earlier, an individual's personality will tell you a lot about how they will operate on a virtual team. Certainly introverts and extroverts will respond differently to working with a team that is not close at hand. What about people with other personality traits?

Try to imagine the good and bad aspects of someone who is highly empathetic. They are wired to care about how other people feel and to do their best to make other people feel better. On a virtual team, they will play a positive role in that they will work to get the team to relate and share personal observations and feelings when they communicate. At the opposite end of the spectrum is a person not attached to other people's feelings, who often will have no idea how others are feeling. People like this are all business, and in communications or meetings will only care about the business at hand. If someone is having a bad day, chances are nonempathetic team members might not notice—and even if they did, they probably wouldn't care to discuss it. These aren't bad people; they just have less empathy for the people with whom they do business. On a vTeam, these types are good in one very specific way—they get the business done—and terrible in another—they don't make much of an effort to know coworkers' kids' names or when a person's birthday might be.

There aren't necessarily "good" and "bad" traits, but some traits certainly work better in specific situations that crop up when you work on a vTeam. It is clearly unwise to assemble people and have no idea what kind of personality traits you might be managing, and to give no thought as to how they might interrelate.

In addition to skill set and personality, a member's working style should be considered when assembling a vTeam. There are people who are diligent about getting homework done on time, and there are people who need lots of supervision and accountability. Unless you want to spend lots of time checking up on people, you might want to do a little research into how well team members can get their work done on their own. It doesn't take me very long to figure out a person's ability to discipline him/herself; I normally give them a few tasks with deadlines and I see clear evidence one way or the other.

Over the last couple of years I have been taking on interns from a local college to help them get some real-life experience in the business world and help our company get a little work done at the same time. All of these interns work remotely, so I will meet with them once or twice, get them oriented to what we are doing, and then assign them some tasks. This is much like the formation of any vTeam. What happens almost immediately is I find in any given group of four students that one person is perfect at getting great work done and right on time. One person does pretty good work and generally hits the deadlines. The other two people do average work at best, and never get it in on time. Performance has nothing to do with their grades—it is just the working style they have adopted. Because I do not have the time or will to change their working style, I just focus on the two who can actually work on their own.

ASSIGNING WORK TO THE vTEAM

As strange as this sounds, there are actually ways to assign work to a vTeam that differ from how you would assign work to an office-based team. Because a remote team cannot see each other work, and because you do not have as much of a chance to pat people on the back and give them kudos, it is good idea to break a vTeam's work up into smaller pieces with shorter deadlines. By doing this, you can identify more quickly those who are behind or who miss their deadlines; at the same time, the rest of the team will have this information as well. Breaking up work into smaller bits also gives the leader more of a chance to compliment vTeam members—and as I discuss in the section called "Rewarding the vTeam," this is critical.

The other difference in assigning tasks to a vTeam versus a traditional team is to be very conscious of whether you give work to a single person or spread it out across people. When a project is given to team members who all work in the same office, the conversations and monitoring are only a few steps away. Give the same project to a group of people who do not work in the same place, and they will often struggle to get their respective pieces together. For this reason, the leaders of vTeams need to consider giving full projects to one team member or breaking the projects up among a small number of members who have already proven they can work together.

COMMUNICATION RHYTHM

One of the most critical factors in making a vTeam work well is communication. Although every team needs communication, a vTeam needs great communication, delivered through the proper electronic means and in predictable rhythms. When you do not see people every day, it becomes important for you to know how and when you will communicate with them next. This is especially true when you have extroverts on the team who derive energy from communication with others.

A simple example is the weekly Monday meeting that a team might hold to discuss the previous week's progress and the current week's agenda. Some team members will look forward to this meeting and find it one of the most important parts of their week. Other team members will dread this meeting because they find it inefficient and slow. It just depends on each team member's personality.

What will be equally important to everyone, however, is that there is a predictable model for how and when communication happens. This is the vTeam's "communication rhythm": its daily emails, phone calls, and texts; weekly conference calls; monthly online meetings; and semiannual in-person meetings. This kind of predictable rhythm of talking helps people feel comfortable and confident that there is a time for them to be heard and to ask questions. Without this kind of rhythm, people fall into states of fear and uneasiness and they waste lots of energy. If you are the leader of a vTeam, state clearly what the regular routine is for communicating, as well as what the proper technologies are for each type of communication. You will be amazed by the productivity gain.

vTEAMS AND THE TUCKMAN MODEL

Another factor that can help you succeed in leading a vTeam is the teamwork concept developed by Bruce Tuckman in 1965. Tuckman talked about the stages every team goes through when it is being developed, whether a sports team, a business team, or a team of people at church. The stages are *forming*, *storming*, *norming*, and *performing*. Let's go through a quick rundown of the concept.

The first step, *forming*, is the formation of the team. The team quickly moves into the second stage, *storming*, in which team members struggle to assert dominance, form relationships, and find their places in the team's overall strategy. After a time, during the *norming* stage, they begin to sort out who is going to play what roles and how everyone can relate to each other. Battles are won and lost, and the team can get to at least 60 percent productivity because its members' routines and tasks have normalized. With effort (and a bit of luck), the team eventually refines its efforts to reach the *performing* stage, in which it is 100 percent effective and wasting no time on politics or confusion over who plays what roles.

As I said, the Tuckman model holds a truth that every team experiences. The trick is to get through the stages as quickly as possible. A basketball team is a good example of this model: you can put five tremendous athletes on the same team and they will be beaten by a team that can cooperate and play together rather than as individuals. The great coaches can get their players to let go of personal agendas and play team ball—to perform at a high level as a group. This is what a vTeam leader must learn to do.

A few specific dynamics cause the Tuckman model to be critical for a vTeam. The first is that a vTeam has a very short time to get to the norming stage before people will just quit working for you. They will sit at their remote sites and get bitter about events that happened in the storming stage, and if a leader cannot get the team members to very quickly trust each other and move on, the team will likewise become dysfunctional quickly. This is a dynamic unique to vTeams; people who share an office building can better sort out issues, talk among themselves, and push themselves at least to the norming stage just so they can be comfortable at work. People who work remotely from each other do not have the same ease of sorting out early team issues. And because they are remote, the leader often does not find out about the problems until lots of resources have been burned.

vTeams can also suffer more acutely from a cycle called the "storming/norming cycle." This cycle occurs when the team achieves norming just before a team member changes position or leaves, throwing the team back into the storming mode—in which people again must sort out their roles. Because vTeams tend to change out people a bit faster than do regular teams, there is a real danger that this constant back and forth between storming and norming will rob the team of its ability to be productive. vTeam leaders must see their role as driving the team to the performing stage and holding it there as long as possible. When performance is disrupted and drops back a stage, the leader must move quickly to address the problem.

> **The Storming/Norming Cycle:** A situation that occurs when a team is continually taken back and forth between a stable state ("norming") to a mode in which members are sorting out their functions within the team ("storming") because of changes within the team.

REWARDING THE vTEAM

Much has been written about how to reward employees. It is well known that money is often third on the list of rewards employees want—recognition for the work performed is number one, and number two is generally the need to feel safe. Because of these facts, vTeams have an inherent problem; it is harder to recognize good work when you are not there to give the recognition in person. It is also harder to have the team recognize one of its members when they are all in different places. A strong vTeam leader ties public recognition to small, easy-to-complete tasks in order to combat this. It is also critical to go out of your way to recognize accomplishments in front of the team, be that on a conference call or in an email. This means a lot more than a direct compliment that no one else hears.

When a leader has the whole team in the office, there are many chances for quick pats on the back and public recognitions. When people do not work in the same office, there are fewer chances to reward someone's hard work. You must go out of your way to give people positive feedback often, and in front of others.

vTEAM TECHNOLOGIES

The tremendous present-day ability to communicate in real time, and virtually free of charge, has enabled people to work together even when they are physically separated. I could write fifty pages of thoughts on how best to use technology to support a virtual team, but I don't want to bore you so I'll keep this succinct. Technology must be applied to facilitate three important things on a vTeam—communication, data aggregation and sharing, and creative thinking.

Communication

Because there is a need to communicate in ways other than in person, and because that communicating needs to be almost as easy as if it were in person, it is critical to design a communication platform that absolutely rocks. Every good communication plan has two components: first, a precise list of what technologies will support the communicating, and second, the agreed-upon set of guidelines for how and when to use communication tools. In the best-case scenario, organizations would use all the standard tools like email, phone calls, and instant messaging, plus virtual conferencing and collaboration tools like Skype, WebEx, and NetMeeting. Then the leader must dictate to the team how and when the tools would be used so that no one wastes time by either failing to use an important tool or over-using an ineffective tool. I've seen predictable variations in the use of communication tools that result from the different ways each one is perceived by the different generations. I frequently see the younger generation use text-based tools when a phone call would be more appropriate, and I have seen older people use the phone when a text option would have been better.

Data Aggregation and Sharing

Data aggregation and sharing is a specific strength of technology. We have spent the last thirty years building software applications that allow people to share data and information. Because a vTeam cannot share data by handing it over the cubicle wall, it is critical for the leader to ensure that the team has the most functional software applications available to support its mission. A ton of collaboration applications are available to help remote teams share data, files, and analytics. Many industry-specific applications are now web-based so that any-one, anywhere can run them at the same time as other users. As a vTeam leader, I want my team to have a shared platform that facilitates group calendaring, archiving files, reporting, and project management. With these tools, each team member can easily stay abreast of what the other members are doing, just as if they were in the office.

Creative Thinking

Creative thinking can be supported with technology through the use of group whiteboarding over the Internet. Certain software applications let teams develop idea trees collabora-tively over the web. Ideas can be captured in databases so they never get lost. Every day I see new web-based tools that allow a group to design, write documents, or brainstorm together. The trick is for the leader to pick the best three or four tools for the specific vTeam to use.

I suspect that just about everyone who reads this book is dealing with some variation on a virtual team; it's not a brand-new concept. What is new is the awareness and availability of specific practices that can greatly improve the productivity of a vTeam. People have spent a lot more energy figuring out how to build technological tools to support vTeams than figuring out what human-based best practices need to be applied. The concept of virtual teams grew on us quickly, and it does have a lot of good aspects that help both individuals and the organizations that employ them. I hope this chapter has inspired you to take a fresh look at how you are operating with your remote team members.

CHAPTER 19

UNLEASHING THE TEAM

Once the game is over, the king and the pawn go back in the same box.

—Italian proverb

NOTHING—LET ME REPEAT THAT, NOTHING—IS A better indicator of the success or failure of an organization than the ability of the team members to perform. In addition, they have to have the will to succeed. Talent without the will to succeed will be wasted. All of my preaching about velocity is not intended to tell you to just go faster. The goal is to build an organization that can move at the same velocity as the market, or maybe even a bit faster. The speed of industries varies; some organizations must move faster than others. Schools and governments generally do not need to move as fast as technology companies do, so there are different expectations for the pace of progress. But remember that the pace of an industry may speed up or slow down depending on the unique factors in it. Although schools generally move at a slower pace than a tech company moves, circumstances may dictate that educational institutions dramatically speed up their rate of change if they want to stay relevant.

This chapter is about unleashing your team, and when I say that, I am talking about instilling a culture that allows—and promotes—the employees' ability to deliver the maximum of their potential to the organization. In my experience, the only two reasons why employees were not giving their best was that they either made a semiconscious choice not to or they felt restricted or distracted in some way by the entity they worked for. In order to move with velocity, you need every member of your team to move in a frictionless way toward completing tasks. To the extent they are held back by lack of understanding of the mission, angst about their manager, lack of incentive, or a generally unhealthy culture, the overall organization suffers. A team member who feels thwarted in bringing his or her best will be personally and professionally unhappy in any position. I guess that is why so many people right now are saying they would love to change jobs! As a leader, you must take the responsibility of unleashing your team members' potential—for their good, and for yours.

I am not recommending that you unleash your team to move so fast they take careless risks and crash into realities that cost you dearly. The goal is not to unleash them into a state of chaos. You want to unleash them so they are not wasting time, talents, and resources in unproductive areas. This will help your people have the ability to innovate and lead your market. If you are not leading, you must be following—and your team will not enjoy learning what to do in the market by watching what works for your competitors. At the moment you rely on your competitors to lead, you're probably too far behind to survive the climb back up.

I am an athlete and have played on many sports teams in my life. I also enjoy following lots of different sports on TV. Some people might enjoy just watching awesome plays, but I love watching how teams function. I happen to especially like soccer, a sport I have played since I was eleven years old. Soccer, much like football or baseball, is made up of a group of people who have to function in harmony in order to win. If any one player fails to do his/her job on any given play, the team can lose. I have played on teams that were completely in sync, and I have played on very talented teams that were a mess.

Sports teams, unlike most organizations, get to test their strategies in short, quick bursts of an hour or two. They get to see clear, concise wins and losses, and they either achieve the championship each year or they do not. Each time, they get to come back a few months later and try to win it all over again. It would be great if organizations could play like this. But organizations have to build strategies and teams over several years, and then wait additional years to find out if they "win." And then, even if they do, they have to show up the next day and keep right on going with no off-season during which to rest.

The best sports teams also have a deep, collective drive to win. This same drive is the special ingredient that can help a less talented and underpaid team rise up and beat the talented and wealthy competition. In fact, this happens with a few teams in just about every sport in every season. This all-consuming will to win helps them work a lot harder, not give up when the game gets tough, and remain selfless in submitting to the team goal of winning.

How sweet it would be if we could get our organizations to have this same passion to succeed, with every single team member giving 100 percent every moment of the day. It *is* possible, of course; a few organizations here and there build that kind of *esprit de corps*. But sadly, it isn't common. I once polled about twenty members of a company I had just developed a relationship with and asked them to anonymously write down on a piece of paper what percentage of their efforts they were really giving to the company. I figured the average would be around 80 percent. When they turned in the scraps of paper, the average was about 40 percent. I was stunned—I figured everyone had the same drive to perform that I had. They all had their reasons, and I went around the room and heard many of them. The experience forever changed my view of the actual effort people often put into their careers.

A powerful leader must have the ability to facilitate a culture that unleashes team members and makes them want to give 100 percent, to strive to give everything they can. All employees dedicate a certain percentage of their "possible effort" to getting their work done. The majority of workers will tell you they rarely give 100 percent. They give enough effort to get paid, or enough effort to look good, or enough to get the job done. But we want them to voluntarily give us 100 percent of their effort, even if they don't necessarily have to give it. This is called their "discretionary effort." That discretionary effort might be an additional 20 percent over what they are giving now, and 20 percent more effort can have a huge impact on the bottom line when accumulated across a large team. Obviously, people only choose to give this discretionary effort when they feel there is a good reason to do so.

Discretionary Effort: The extra effort put forth by employees when they are motivated or inspired to give a full 100 percent—after they have given a lesser number traditionally.

This is especially true when organizations have been introducing wave after wave of new technology into the operation. Many employee bases across the United States are collectively exhausted from a decade of changes in the marketplace and repeated additions of new digital plumbing they must manage as part of their jobs. As businesses strive to raise the amount of revenue per employee and leverage technology to become more efficient and productive, we have also broken down relationships and put a premium on people staring at a screen all day. Although this drives profits and productivity, it does not do much for the human element. Technology has just as many downsides as it does upsides, so we need to take very conscious steps to improve how we go about getting things done.

Let's look at some specific ways in which you can go about encouraging the discretionary effort that is pent up in most organizations. These practical investments you make in your team will yield much progress, and I use the word *investment* because each of them takes a commitment of time and, in some cases, money.

REWARDING RIGHT-BRAIN SKILLS

One of the most important cultural changes organizations can make is to institute a reward system for right-brain skills. Daniel Pink's book *A Whole New Mind* points out how valuable right-brain skills are becoming. I absolutely agree. I have been watching computers grow in their ability to do the left-brain tasks that have long been done by human beings. In a practical sense, this means tasks like financial auditing, reviewing insurance claims, processing tax returns, and the like can be done better by computers than by humans. Now that we have learned how to build sophisticated layers of rules, it is simply a matter of mining the expert's brain and converting those ideas to code. In short, computers are better at performing logical tasks and will continue to replace humans in doing repetitive, noncreative work. However, what computers will never be able to do is be creative, innovative, artistic, or unpredictable by themselves. Human beings are unique in our ability to provide these right-brain skills.

For this reason, creativity and innovation are going to be extremely valuable going forward. A leader who wants to win in the future must become very good at inspiring these traits. Witness the Google policy of demanding almost all of their employees to dedicate 20 percent of their time to "R&D thinking."

If you are not convinced about this right-brain value concept, then allow me to come at it from another direction. In a world that is gaining more and more people and in which global competition is increasing tremendously, the ability to find new products to offer and new ways of doing things is going to be at a premium. If you provide a product or service that is not unique, it will quickly become commoditized, and you will be stuck waging price wars you'll struggle to win. Of course, the Internet will allow people from anywhere in the world to compete with you, so good luck thinking you can hold on to a local customer base as your way out. The only way you can separate yourself from all of the global competition is by being clever, innovative, and creative about how you deliver your product or service. Being boring and unchanged will result in you losing market share to those who provide new, more valuable options. This is especially true if you are a banker, lawyer, CPA, doctor, insurance broker, real estate broker, etc.; these traditional industries will suffer in the next decade unless they continue to retool their operating models.

I get tired of reading mission statements that have the words *creativity* and *innovation* in them when I know damn well the company has done nothing new in years. It is hard to be innovative in a world with so much competition. It is hard to be creative when just operating the organization is more than a full-time job. It is hard to get your mind off your To Do list and think big thoughts from a fifty-thousand-foot perspective. Regardless of these demands, you should either live what you say you are in your mission statement or take the inaccurate parts out.

RANKS-OFF MEETINGS

This concept is borrowed from the military, and it is a brilliant way to encourage velocity. In a "ranks-off" meeting, everyone has an equal say in the discussion, and everyone must win any debate by the power of their ideas and their knowledge. In this type of meeting, position, tenure, and rank have nothing to do with how decisions are made. If you want ideas to flow or need to get plenty of input before making an important decision, a ranks-off meeting will get you there faster than a regular meeting will.

In order to run a meeting like this and get the desired impact, the traditional leaders must step aside and allow others to have a voice. They must allow debate and take input from everyone equally. The moment a high-ranking person tries to win a debated point by using rank instead of ideas, the spirit will be lost. When run correctly, people will begin to crave having the ranks-off designation attached to meetings because it frees everyone to give their input and have their ideas heard. It also creates velocity because it enables people to get their ideas into play quickly and encourages a nimble, fast-moving discussion.

Only by designating specific discussions as ranks-off and giving everyone the ability to win the group over with knowledge and good ideas can we fully leverage the whole staff. You will reach goals more quickly and easily because you will not only have mined all the talent you have; you will also have created buy-in from all involved because they had their chance to give input. Do this, and you will have an engaged team that will move with velocity; don't do this, and you will suffer from the friction of the unspoken word and the unshared idea.

LAUNCHING AD HOC PROJECT TEAMS (TIGER TEAMS)

One of the main inhibitors of velocity is the inertia that grows within an organization when people have been in the same jobs, on the same teams, doing things the same way for far too long. For decades, we have built structures that contain certain routines and rhythms, and we have now become pretty wedded to those habits. This holds true for churches, hospitals, government entities, businesses, and schools. We form these structures rigidly and have a hard time imagining them in any different configuration than the one we're used to. The problem with this is that the world is going through huge changes as it experiences macro trends like globalization, technology proliferation, generational differences, and new communication patterns. These trends force us to take a different look at how we organize institutions and identify ways to create fresh viewpoints on how we could operate.

Tiger Team: A small, diverse, ad hoc team assembled to address a specific issue.

One way to shatter the inertia of existing structures is to assemble small ad hoc teams to solve specific challenges— "tiger teams," as I call them. The trick here is the makeup of the team. It is critical to assemble young people with a fresh eye, older people with years of experience in the industry, and maybe an outside consultant who is not married to the way you have always done things. The mistake many leaders make today is that they see a problem, become frustrated by it, and yet seem to believe it can be solved if they just think hard enough. The fallacy in this is that you can think as hard as you want, but if you can't see new possibilities, you will just think really hard about the models you already have in place. Creating an ad hoc tiger team that has no vested interest in doing things the old way makes a lot of sense. What makes even more sense is to fill that team with people who have awesome right-brain skills!

The person most often left off a tiger team, but who definitely should be included, is the intern. You heard right: the intern. Take the youngest person around, the one who in a way is just visiting, and get him or her involved in your most important discussions. Let an intern participate on the thorniest issues. What you will find is that even though interns often are intimidated by the age and experience in the room, if you give them enough time to adjust and force them to share their opinions, you will often be shocked at what they can add to a discussion.

Another wild card on the tiger team is the programmer. Experiment with adding a skilled programmer to your business discussions and notice the input you get. Because programmers are artistic in nature, and because their minds see things in a unique fashion, they make great observations as to how problems can be solved.

Don't waste the assets you have within your company by getting stuck in the routine of relying solely on your executive team to develop solutions. Reach out to the corners of the organization you have not engaged before. In addition to generating some great ideas, you will be sending a powerful message to your employees: you value all opinions, even those on the fringes.

Increasingly, I have been leveraging tiger teams in the organizations I work with. When we identify a specific task—for example, trying to figure out how to use technology to lower customer turnover rates—we pull together cross-functional, cross-generational teams to address what might be possible. We find that the team is often energized simply because of the eclectic combination of talents. Then they are excited to succeed when they develop intriguing new concepts for solving the problem. If you want new solutions to old problems, assemble a new combination of viewpoints to tackle the problem. This seems simple, but many leaders fail to do this because they assemble teams of people much like they themselves are, and they end up with little diversity in their teams.

CIVILIAN TECHNOLOGY TRAINING

Training is an investment. It takes time away from the core tasks people must perform, and the returns can often be hard to measure. Yet if you want to truly create a powerful team that operates tightly within a healthy culture, training is key. I am not just talking about training people how to use Microsoft

Office, either—you can get that free at Microsoft's website. I am talking about training the staff in areas like the logic behind the organization's technology strategy. Teaching them why the digital plumbing is organized in the way it is and, specifically, what their role could be in improving it. Training them on software applications that can be used to mine important data from the databases. Training them on the basic foundations of technology like security, backups, and file structures. Further, we should train on more complex things such as how the Internet works, technology vocabulary, and how software applications are built into the organization.

At this point, some of you may be wondering why you should bother teaching these things at all. How does it make any sense to take people out of their jobs for hours at a time to teach them this level of technology? Let me answer by way of analogy. Safety training and sexual harassment training are forms of employee education large organizations conduct without thinking. They invest in them because failure on the employee's part to make good decisions in either of these areas can cause the company a whole lot of money in lawsuits (although I am sure the organization also cares about their employees and does not want them to get hurt or be harassed).

We have swamped our operations with technology and then, in many cases, done *very* little training. The lack of training translates into people not being able to take advantage of the investments organizations have made in these tools. We do a paltry amount of initial training, get people doing just the basics, and then call it good. We somehow expect people to pick up the rest on their own, or we seem to not care if they learn at all. The potential result is millions of dollars lost. This is no exaggeration. I see organizations every day spend millions of dollars on software that in reality is only 50 percent utilized.

On a micro scale, my experience has shown me that most users tap only 30 percent of the features in Microsoft Word. Multiply that 70 percent loss of functionality by the tens or hundreds of employees who use the software systems that run your company and you'll see what I mean.

There are other, secondary benefits to providing deep technology training. Your people will not only respond by using the tools you have provided in better ways; they will also feel better about themselves. They will appreciate that you are making this kind of investment in them. I have seen companies institute simple monthly lunch-and-learn programs in which they buy pizza for the staff and then train for an hour on a new topic—that's a great ROI for the cost of the pizza. Many people show up for both the free meal and the chance to learn, and they really appreciate both. They take their new skills right back up to the office and use them to be more effective.

GREEN LIGHT TECHNOLOGY PLANNING

This section describes a process of holding cross-generational and cross-functional brainstorming sessions on technology strategies and possible improvements to the digital plumbing. Some of you might ask, "What does technology planning have to do with the culture of an organization?" Good question. The answer is that many employees spend a high percentage of their time using technology at work. Improvements in the digital plumbing not only help the productivity of the organization but also improve the quality of life for employees. And improvements in quality of life tend to have a very positive impact on employees' love of the organization. For this reason, it is a wonderful idea to develop a process where you gather cross-functional teams to discuss the technology strategy currently in place.

During the dot-com era, there was one start-up that would meet once every two weeks and hand out an award for the worst mistake made over the preceding weeks. Within their culture, their CEO developed an environment where it was OK for people to share what they have done wrong so that others could both learn from this mistake and also know it was alright to at least try things and fail. In many cases, the team had a pretty good idea who was going to win at each session because they had already heard through the grapevine about a large blunder. The meeting was always filled with laughter, yet everyone would get serious when talking about how to avoid the same mistakes in the future.

I love this concept and think it should be applied directly to how we ask people to use technology. At what I call our "Green Light Technology Meetings," we ask, "What is the biggest mistake we are making with our technology systems?" The answers we get are often enlightening and valuable. Here is a sample of things I have heard in these meetings:

- We created a web-based system so we could do our work from home; then, when the ice storm came, I was docked pay for not coming in to the office, even though I did everything I normally do from my house.

- My boss makes me print out invoices and store them in a file cabinet, even though the new system does a great job of making them available on my screen and the backups are bulletproof. We *never* go back to that paper.

- My department at the bank has no ability to see customer data from the other departments, so when a customer asks me a question about a loan, I have to refer them to someone down the hall.

- I had a patient die last month because the information the doctor provided in writing was so illegible that we made a mistake. I am not even allowed to say this out loud, but I am so upset that I just have to ask why we have not computerized that part of the process.

- I spend millions of dollars on marketing and have no way to track any actual sales that come from my investment. There is no technical reason why I should not be able to do this. We just do not track enough customer data to be able to monitor the impact on a specific group I have marketed to.

- I have lots of ideas, but I am not even going to tell them to you—our IT department has no ability to get any of them done.

The saddest thing I have seen in getting civilians to open up and do some green light thinking on how technology can be improved is their apathy toward making these observations. Many of them consider the IT department to be so swamped already that they would not dare raise these issues. In some cases, they have actually tried to petition the IT people to solve specific problems, only to be met with a barrage of technobabble that ensured they would never ask again. The problem is that we are allowing the technical people to make the decisions on what is most important. When we do that, we often get a ton of time spent on upgrading the servers to new operating systems, and no time spent making sure that patients don't die.

GEEK SEEDING

One of the major mistakes lots of organizations make is clustering all the smart technology people in a dark room—separated from any other operating group as if they were lepers or something. Of course, lots of tech people are introverts who like relating to a machine, so they willingly hide away. More progressive organizations are learning, however, of the huge benefits inherent in moving IT people and tech-savvy employees out into the operating groups so they can work side by side with employees who are experts in their own departments—a situation that helps IT professionals apply tech in better ways.

Part of this process is rapidly identifying A players and locking them in with a specific plan for retention. A team member who not only has a specific business skill (accounting, sales, marketing, engineering, etc.) but also has a high level of technology sophistication is extremely valuable. Workers like these are not overly plentiful at the moment, but they do exist in most organizations, and the good news is that their numbers are growing as the next generation hits the workforce. HR departments should be on the lookout for employees with this special blend of talents, and when found, they must be locked in so they don't leave.

There are two reasons to pay special attention to this breed. The first is they can provide huge value wherever they are placed. They add a perspective to the work flow around them that can result in great improvements. They will quickly size up the work to be done and will apply whatever technological skills and tools they have to make things work better. As they show others how to use these tools in their work, they create value by transferring knowledge. This way, even when they move on, the benefit will stay.

The other reason to reward these hardworking geeks well is that they are normally the hardest to hold onto. Technologically knowledgeable A players are imminently portable. They can find work elsewhere and will often be offered more money than you are currently paying them. In some cases, they will learn your business and then go out on their own, using their skills and talents to open their own company. They will assimilate all they have learned from you and add their own magic touch to compete with you in the market. When these employees announce they are leaving your company, you will struggle to hold on to them. You will offer more money and whatever other inducements you can, but it will be too late. Your star has already moved on.

For these reasons, you must practice "geek seeding" in your company. Identify these A players and put them on a special path that takes advantage of the value they bring and keeps them in your employ. They should be put in a special class of employment. They should have access to large amounts of discretionary training. They should get to sit in on important executive meetings from time to time. These are the high-ranking execs of the future. They may not have all the management savvy they need yet, and they may not fully understand the drivers of the operation, but they have the foundation that will be a powerful platform as they grow over the coming decades. Until knowledge and execution are fully in the realm of the computer, we will depend on humans to accomplish organizational goals. Do not make the mistake of churning through your most talented and technology-oriented human assets.

Geek Seeding: The practice of identifying technologically skilled A players, placing them throughout an organization, and developing a plan to retain them.

REVERSE MENTORING

I love the concept of reverse mentoring and first ran into it while working with a client. This organization had developed a process by which young, tech-savvy employees mentored the fifty- to sixty-year-old leaders on how to use technological tools. This is a fast and inexpensive way to get the wise elders in the organization to understand how they can leverage their experience and be more productive with technology.

One awesome side benefit of reverse mentoring is the relationship building that happens naturally when someone with years of wisdom and experience befriends someone who is just starting out. I cannot emphasize enough how valuable it can be to a team when a young, tech-savvy person merges skills with an older, market-savvy person.

As an entrepreneur who has built multiple companies from scratch, I have seen firsthand the importance of a team that consists of quality people, working in a healthy and vibrant culture, committed to the success of an endeavor. As a consultant, I see many clients who reinforce this observation either because they get it or because they don't. Motivating a team is not enough—inspiration can be temporary. You must create a culture that makes your team want to give you its discretionary effort. People want to know for sure that they are making a difference, that they are important to the success of the organization. Christmas parties, casual Fridays, and floating personal days do not get you there. A culture that fosters achievement, provides the right tools, and then recognizes contributions will.

In this section, I have discussed a few skills and processes you will need to build a high-velocity culture. You will have to bridge the technology gap between individuals in different generations. You will have to institute enterprise social technologies, and learn to lead virtual teams. Above all else, providing advanced technology, top-notch training, and a clear mission is a good place to start. Cultures can be altered. Cultures can be created. Company culture is built by leaders (who set the tone for it) and team members (who self-create the values that make it up). Cultures set the expectations for new employees. Cultures define how attached employees are to success, and how fast and hard they will work. Cultures are living, breathing representations of a team's collective beliefs and will.

Creating a culture that facilitates innovation at every level is now mandatory. This is because technology has unleashed employees to search, learn, and experience much more than ever before. This gives every employee the tools to find the best practices for their position and then translate these into improvements for the whole team. Technology and a velocity-based culture provide an infrastructure in which employees are encouraged to be creative about improving their tasks—and the will to implement them. When you support a high-velocity culture with an inspired high-beam strategy, you might actually get the discretionary effort that could make a dramatic difference in your chance for success. Fail to create a high-velocity culture, great digital plumbing, and a high-beam vision, and the organization will slowly perish as you hemorrhage A players who leave for greener pastures—often out of boredom and lack of a mission that inspires them.

CHAPTER 20

GOING FORWARD . . .

Life is either a daring adventure or nothing.
Security does not exist in nature, nor do the children
of men as a whole experience it. Avoiding danger
is no safer in the long run than exposure.

—Helen Keller

MY MOTHER TAUGHT ME MANY YEARS AGO THAT there is a thing called the Truth with a capital T. This is an absolute. There is a Truth to how the future will unfold, and none of us is insightful enough to know exactly what it will look like. There is a Truth about how data flows through computers. There is a Truth to the most productive and healthy ways to run a team of people. Unfortunately, we *all* have flawed perspectives in trying to grasp that Truth. Some perspectives are just more flawed than others.

Our perspectives are flawed because we often find ourselves looking through old filters. They can be flawed by beliefs that are just plain wrong, or they can be incomplete because we simply lack the necessary information. Leaders often make decisions about technology when they know next to nothing about it, but leading with incomplete or flawed perspectives means we make poor decisions and lead others down painful paths.

The pace of change in our world demands that we constantly revise our perspectives. If we do not, we will be looking at a new world through old eyes. You have to see clearly in order to go fast. That's what the velocity manifesto is all about.

It is difficult to change your perspectives. The longer we work from one set of perspectives, the harder it is to change. The more successful we have been, the harder still. Yet the world is littered with stories of people who were at one time successful but were never able to duplicate the feat. This book will be valuable for a few years, and then there will be a need for another set of perspectives to build on. And so it will go forever. If you take nothing else away from this book, please understand that there is not a fixed inventory of leadership skills. The skills that worked twenty years ago may only be part of the story now, and if you decide somewhere along the way you have all the leadership skills you'll ever need, you just broke the first rule in being a lifelong learner.

If you are a leader, don't just hang out until retirement. Fix your digital plumbing, develop a high-beam strategy, and build a high-velocity culture. Rely on the velocity manifesto as you lead your organization to unimagined success. Keep pace in this new era; don't be a victim of stale thinking and a fatal reliance on the way you have done things in the past. Frankly, such an attitude is a bit lazy, and it will not serve you as a leader in a technology-augmented world!

Get in touch with me if I can be of any assistance. The great thing about a wired world is that without having any idea where I am at this moment, you can still reach out to me. The futurepointofview.com website provides an easy way to connect, and you can also reach me through www.velocitymanifesto.com.

ABOUT THE AUTHOR

Scott Klososky has founded numerous successful organizations, including two (Paragraph and Webcasts.com) that were sold for more than $200 million in aggregate. His unique insights into technology, business culture, and the future of business allows him to travel the globe as a speaker and consultant, advising senior executives in organizations ranging from Fortune 500 corporations to universities and nonprofits on topics such as IT strategy, social technologies, and leadership. He has previously worked with numerous companies—Cisco, Newell Rubbermaid, Lockheed Martin, Ebay, Volvo, The Hartford Financial Services Group, Great Clips, and Marriott, among others—as well as associations—the International Franchise Association, the Korean Ministry of Information, the Mortgage Bankers Association, the American Payroll Association, and the Association of Equipment Manufacturers.

Scott currently serves on the board of Alkami Technology, his latest technology start-up. Alkami is redefining online banking by creating a platform that allows banks and credit unions to deliver customized services to demographics like kids, senior citizens, and young married couples.

Scott publishes a blog at www.technologystory.com, and he can be found on Twitter at @sklososky.